weather
for the mariner

SECOND EDITION

weather
for the mariner

WILLIAM J. KOTSCH, REAR ADMIRAL, U.S. NAVY (RETIRED)

NAVAL INSTITUTE PRESS   ANNAPOLIS, MARYLAND

contents

CHIEF, OFFICE OF BOATING SAFETY
UNITED STATES COAST GUARD
WASHINGTON, D. C. 20590

FOREWORD

With the advent of the Federal Boat Safety Act of 1971 and the creation of the Coast Guard's Office of Boating Safety, there is an ever
increasing awareness of the importance of recreational boating safety.
Recreational boats have increased enormously in number over the last
few years with more and more novice boaters taking to the water.

The Coast Guard and the Coast Guard Auxiliary have prepared the
Weather Advanced Specialty Course to instill in the boater an
appreciation and understanding of available weather information.
This text is one in a series developed by the U.S. Naval Institute
for use in the Coast Guard Auxiliary Membership Training Program.
Although designed to make the boater more aware and sensitive to
the ever changing environmental conditions, this text will not make
professional weather forecasters of those who undertake the course.

While our primary concern is to instruct individuals in recognizing
adverse changes in the weather, this text will also serve to enhance
the boaters' appreciation of the total boating environment in good
weather as well as bad. An understanding of the weather will add
enjoyment to your days of safe boating.

D. F. LAUTH
Rear Admiral, U. S. Coast Guard
Chief, Office of Boating Safety

preface to the second edition

As every student of the earth's environment knows, continuous progress has been made in the sciences of meteorology and oceanography since this book was first published seven years ago. Some of this progress includes high-speed computer modeling, the interpretation of weather satellite data, the refinement of weather and oceanographic forecasting techniques, the implanting of automatic, experimental, environmental data buoys in the Atlantic and Pacific oceans, and so forth.

In this revision, some newer ideas and practices have been included, and changes in the material were necessitated by organizational changes within the federal government, operational changes within federal agencies and the Department of Defense, changes of radio frequencies, the creation of the National Oceanic and Atmospheric Administration (NOAA), and many other factors. Innumerable paragraphs required revision and updating. In some cases, entire sections have been re-written, and in one case, an entire chapter required re-writing. Throughout, however, the general method and order of treatment of subjects and material have remained essentially the same—uncomplicated, nonmathematical, and hopefully, interesting.

With the trend of the United States toward "metrication," some additional material on metric units and metric conversions has been included. And it is suggested that mariners and weekend sailors begin to "think metric" now. New sections on the sea-breeze and land-breeze phenomena and equivalent chill temperatures have been included. A total of 54 new, or extensively revised, diagrams and drawings and 5 additional tables have been included in this revision. A great deal of material has required up-dating because of changes (both operational and organizational) which have occurred in the last seven years. Additions have been made to Appendix A and Appendix B, and a new Appendix C, containing the latest and now permanent worldwide lists of hurricane and typhoon names, has been added.

The author is extremely grateful to the following individuals for their generous cooperation, helpful suggestions, or permission to use material: Dr. Robert M. White (Administrator of the National Oceanic and Atmospheric Administration (NOAA)), Dr. J. R. H. Noble (Assistant Deputy Minister, Atmospheric Environment Service of Canada), Dr. George P. Cressman (Director, National Weather Service (NOAA)), Dr. David S. Johnson (Director, National Environmental Satellite Service (NOAA)), Mr. Stanley B. Eames (Director of Public Affairs, NOAA), Brigadier General John W. Collens, III, USAF (Commander, Air Weather Service (MAC)), and to Lieutenant Colonel Harold E. Headlee, USAF and Technical Sergeant Edward W. Bunyea, USAF, of the Organization of the Joint Chiefs of Staff.

And last, but assuredly not least, the author is deeply indebted to Professors Sverre Petterssen and Thomas F. Malone who, 36 years ago at the Massachusetts Institute of Technology, compelled the author (by kindly threats and menacing persuasion) to take *detailed* notes during class lectures and in the laboratory. Some of these notes are as valid today as they were over three decades ago, and have served as the basis for some of the material contained in this revision.

William J. Kotsch
Rear Admiral, U. S. Navy

preface to the first edition

The aim of this book is to present in an uncomplicated and readable manner the basic principles of modern meteorology and certain practical aspects of the newly emergent science of oceanography. Intended for individuals with a love for boats, motorboats, and yachts, and with little or no previous acquaintance with the subjects of meteorology and oceanography, the book is written in a nontechnical vein. The use of calculus and high-level theory has been deliberately avoided. Instead, the book is short, nonmathematical, and contains many illustrations—to provide pleasurable as well as informative reading.

The physical phenomena of the earth's envelope of air (atmosphere) are exceedingly numerous, exciting to learn about, and of great importance to all men—particularly to those who are, in one way or another, connected with the sea. All types of small-craft or ship operations are affected to some degree by the state of the air and ocean environments—some much more so than others. Consequently, a knowledge of current weather and sea conditions *and* of forecast conditions is essential for the safe and efficient operation of all sizes of sea-going craft, in coastal waters as well as far out to sea.

Anything that disturbs the equilibrium of the earth's atmosphere is likely to produce some form of danger. The same applies to the oceans. Atmospheric storms and ocean waves, for example, force themselves on the attention of everyone who follows the sea. No seafarer can dare to ignore them. An ignorance of weather signs and sea conditions could easily be the prelude to embarrassment or failure—and perhaps to disaster. It is the author's hope that this book will provide a general knowledge of the most important weather and oceanographic "signs" and that it will help to minimize air and ocean environmental threats by use of this knowledge.

The author is extremely grateful to the U.S. Coast Guard, to members of the staff of the U.S. Naval Institute, and to the several copyright holders and many friends who have kindly made illustrations and other material available for his use.

William J. Kotsch
Captain, U.S. Navy

weather
for the mariner

1

weather warnings and displays

"How is it possible to expect mankind to take advice when they will not so much as heed warnings?"

—J. Swift 1667–1745

Admittedly, it requires more wisdom to profit from good advice than to give it. But things have changed considerably during the past three centuries—especially from the standpoint of weather and sea-condition forecasts, advisories, and warnings.

Today, no self-respecting jet-age aviator would dream of taking off without a thorough knowledge of, and the latest advisories regarding, altimeter settings, icing conditions, winds, clear-air turbulence, the jet stream, and so forth. Similarly, no self-respecting mariner at any echelon should venture from port without an understanding of, and the latest available information concerning, the various types and categories of weather and wind warnings and displays which are especially designed, issued, and exhibited to keep him out of serious difficulty.

Few people are more sensitive to the elements than those who sail the sometimes violent interface between the sea and the atmosphere. And few people have a greater need for regular, accurate, and timely information concerning the current and predicted state of the earth's environment. This information is readily available from the National Weather Service (NWS) of NOAA, The National Oceanic and Atmospheric Administration of the U. S. Department of Commerce. NOAA's National Weather Service regularly provides marine weather and sea-condition reports, forecasts, and warnings for all who "go down to the sea in ships," whether for business or recreation. All sorts of information and advice is available in the form of charts, maps, pamphlets, radio broadcasts, facsimile broadcasts, telephone and radiotelephone services, and so forth. Anyone who does not make good use of these excellent facilities and services is very foolish indeed.

When warnings are issued, don't venture forth unless you are absolutely certain that your boat can be navigated safely under the predicted conditions of wind and sea. If you spot a warning display at a U. S. Coast Guard Station, a marina, a yacht club, or at some coastal point, be extremely cautious. Make certain that you have the appropriate NOAA National Ocean Survey (NOS) charts and other publications aboard which cover your part of the coastal or Great Lakes waters. It's a good idea to check your local office of the National Weather Service or the National Ocean Survey for information on how and where to obtain essential aids to navigation.

**YARDSTICKS
FOR WARNINGS**

The most meaningful and universally accepted basis for describing and classifying the various types of meteorological warnings in coastal waters and at sea is that of *wind speed*. Winds may be associated with middle- or high-latitude weather systems, with closed cyclonic (counter-clockwise in the northern hemisphere) circulations of tropical origin inside or outside the tropics, or with weather systems of tropical origin other than closed cyclonic circulations.

Wind warnings associated with weather systems located *in latitudes outside the tropics,* or with weather systems of tropical origin other than closed cyclonic (rotary) circulations, are expressed in the following way:

Warning terms	*Equivalent wind speeds*
Small Craft Warning	Winds up to 33 knots (38 mph), and used mostly in coastal and inland waters
Gale Warning	Winds of 34–47 knots (39–54 mph)
Storm Warning	Winds of 48 knots (55 mph) or greater

Wind warnings associated with *closed cyclonic* (rotary) *circulations of tropical origin* are expressed in the following way:

Warning terms	*Equivalent wind speeds*
Tropical Depression	Winds up to 33 knots (38 mph)
Tropical Storm	Winds of 34–63 knots (39–74 mph)
Hurricane or Typhoon	Winds of 64 knots (74 mph) or greater

An easy-to-memorize system of pennants, flags, and lights (for nighttime) as shown in figure 1–1, is displayed at most coastal points along the seacoasts of the United States and many other countries when winds dangerous to navigation are predicted for any coastal area. This warning scheme is as follows:

Small Craft Warning—One red pennant displayed by day and a red light over a white light at night indicate that winds up to 33 knots (38 mph) or sea conditions dangerous to small-craft operation are predicted for the area.

Gale Warning—Two red pennants by day and a white light above a red light at night indicate that winds of 34–47 knots (39–54 mph) are forecast for the area.

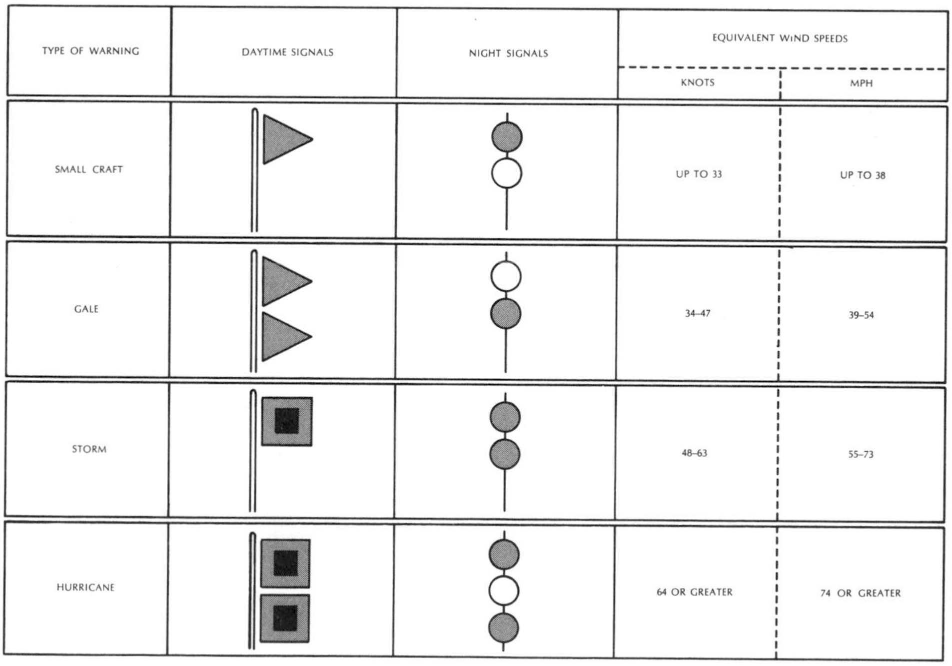

TYPE OF WARNING	DAYTIME SIGNALS	NIGHT SIGNALS	EQUIVALENT WIND SPEEDS	
			KNOTS	MPH
SMALL CRAFT			UP TO 33	UP TO 38
GALE			34–47	39–54
STORM			48–63	55–73
HURRICANE			64 OR GREATER	74 OR GREATER

Figure 1–1. Warning displays.

Storm Warning—A single square red flag with a black center displayed by day and two red lights at night indicate that winds of 48 knots (55 mph) and above are predicted for the area. If the winds are associated with a tropical cyclone (hurricane), storm warnings indicate forecast winds of 48–63 knots (55–73 mph).

Hurricane Warning—Two square red flags with black centers displayed by day and a white light between two red lights at night indicate that winds of 64 knots (74 mph) or higher are forecast for the area. (Displayed only in connection with a hurricane.)

Any individual worth his salt who follows the sea is concerned for the safety and well-being of his passengers and crew as well as himself. It goes without saying that one must be thoroughly familiar with the various warning criteria and what they mean, and one must recognize instantly—*and heed*—the four types of warning displays.

For almost 170 years, wind force (speed) has been conveniently expressed by means of the *Beaufort Scale*, a numerical scale devised in 1808 by Admiral Sir Francis Beaufort of the British Navy. It was based originally on the amount of canvas that a man-of-war of the period could carry with different winds. The numbers on the scale ranged from zero (representing calm conditions) to 12 (representing a hurricane "such that no canvas could withstand"). With the disappearance of sailing ships, the scale as originally designed became unsuitable and has been revised on several occasions.

Today, by international agreement, all wind reports are encoded and plotted on weather maps in knots. This may be changed in the relatively near future to conform to the metric system (i.e., meters per second (mps), kilometers per hour (kph), etc.). From Appendix A, we see that 10 knots = 11.5 mph = 18.5 kph = 5.1 mps. Frequent reference, however, is still made, and will continue to be made, to the Beaufort Scale. For more about the Beaufort Scale, see photographs and Table 11–1 in chapter 11.

BASIC DEFINITIONS

About the middle of the 19th century, one of England's most famous prime ministers, Benjamin Disraeli, stated unequivocally, "I hate definitions!" And he meant it. However, marine activity of any type is antithetical to this kind of attitude. With regard to advisories and warnings relating to weather systems of tropical origin, one *must* understand the terminology and catch phrases used by meteorologists, commentators, and newscasters on radio and television programs, facsimile and radio broadcasts, and automatic telephone answering services in order to realize the significance of the information or the nature of the threat and to take advantage of the warning information. The following few definitions should be at the fingertips of the reader:

Cyclone—A closed atmospheric circulation rotating counterclockwise in the northern hemisphere (clockwise in the southern hemisphere). (See figure 1–2.)

Tropical Cyclone—A warm-core (center warmer than the surrounding air), nonfrontal cyclone of synoptic scale (wave length of approximately 550–1350

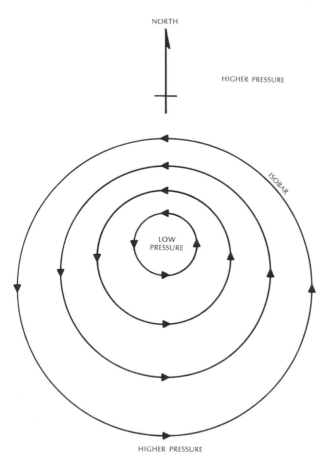

Figure 1–2. Cyclone.

nautical miles), developing over tropical or subtropical waters and having a definite organized circulation. (See figure 1–3.)

Tropical Disturbance—The weakest recognizable stage of a tropical cyclone in which rotary circulation is slight or absent at the earth's surface, but is possibly better developed at higher levels in the atmosphere. There may be one closed surface isobar (line of constant pressure on a weather map) or none at all, and no strong winds. It is usually 100–300 miles in diameter, having maintained its identity for 24 hours or longer. (See figure 1–4.)

Tropical Depression—The weak state of a tropical cyclone with a definite closed circulation at the earth's surface, and one or more closed surface isobars, with wind speeds less than 34 knots (39 mph). The wind speed is a 1-minute mean. (See figure 1–5.)

Tropical Storm—A warm-core tropical cyclone with closed surface isobars and highest wind speeds (1-minute mean) of 34–63 knots (39–73 mph), inclusive. (See figure 1–6.)

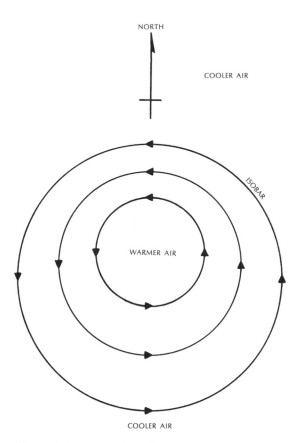

Figure 1–3. Tropical cyclone.

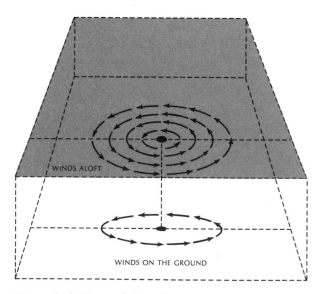

Figure 1–4. Tropical disturbance.

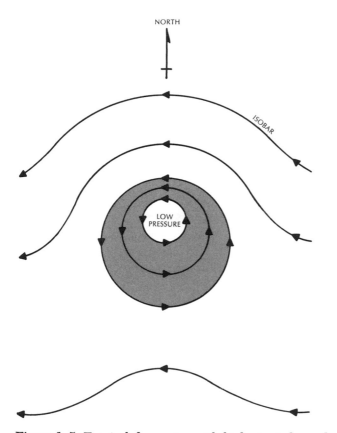

Figure 1–5. Tropical depression with highest wind speeds up to 33 knots (38 mph).

Hurricane–A warm-core tropical cyclone with sustained surface winds (1-minute mean) of 64 knots (74 mph) or higher. (See figure 1–7.)

Hurricane Watch–If a hurricane threatens a coastal or inland region, a "hurricane watch" is established, covering a specified area and indicating the duration. It means that hurricane conditions are a real possibility. It does *not* mean they are imminent. When a hurricane watch is issued, everyone in the area covered by the watch should listen for further advisories and be prepared to act quickly if hurricane warnings are issued.

Hurricane Warning–When hurricane conditions are expected within 24 hours, a "hurricane warning" is issued. These warnings identify coastal areas where winds of at least 64 knots (74 mph) are expected to occur. The warnings usually describe coastal areas where dangerously high water or exceptionally high waves are predicted, even though winds may not quite reach hurricane force. When the warning is issued, every precaution should be taken *immediately*. The warning means that hurricane conditions may be expected within 24 hours. That's not much time to get things done. If the path of the hurricane is unusual or erratic, the warnings

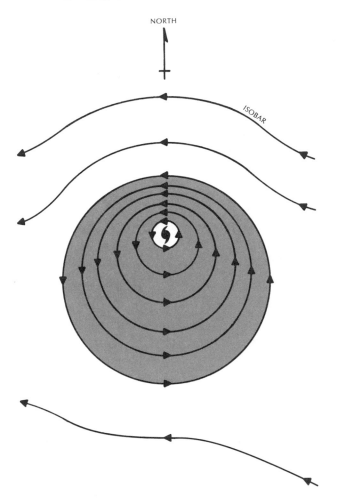

Figure 1–6. Tropical storm with highest wind speeds from 34 to 63 knots (39 to 73 mph).

may be issued only a few hours before the onslaught of the hurricane conditions (high and gusty winds, torrential rain, very poor visibility, exceptionally high tides, perhaps a storm surge, and so forth).

Hurricane Season—That portion of the year having a relatively high incidence of hurricanes. In the North Atlantic, Caribbean, and Gulf of Mexico, it is usually regarded as the period from June through November. In the eastern Pacific it is usually regarded as the period June through November 15th. In the Central Pacific, it is usually regarded as the period from June through October (from 140° west longitude westward to the 180th meridian).

Typhoon—A warm-core tropical cyclone in the western North Pacific (west of the 180th meridian) with sustained surface winds (1-minute mean) of 64 knots (74 mph) or higher. This is the same phenomenon as a hurricane. Typhoons are usually somewhat larger than hurricanes, frequently more intense, and occur more often.

Figure 1–7. Hurricane with sustained surface winds of 64 knots (74 mph) or higher.

Typhoon Season—There is no true typhoon season. Typhoons in the western Pacific can—and do—occur in every month of the year. However, 90 percent of the typhoons occur between early June and late December. A maximum (22.6 percent) of the total occurs in August, and a minimum (0.6 percent) in February.

Quadrant—The 90-degree sector of the storm centered on a designated cardinal point of the compass. An eight-point compass rose is used when referring to quadrants.

Example—The north quadrant refers to the sector of the storm from 315° through 360° to 045°. (See figure 1–8.)

Semicircle—The 180-degree sector of the storm centered on the designated cardinal point of the compass. A four-point compass rose is used when referring to a semicircle.

Example—The south semicircle refers to the segment of the storm from 090° through 180° to 270°. (See figure 1–8.)

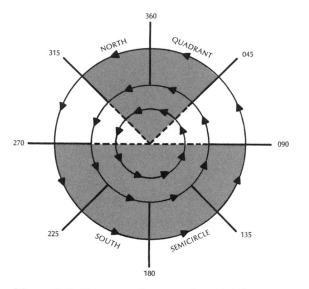

Figure 1–8. Storm quadrants and semicircles.

BROADCASTS AND RELATED INFORMATION

The National Oceanic and Atmospheric Administration (NOAA) was established on 3 October 1970 in the U. S. Department of Commerce as part of the President's environmental reorganization within government. The purpose of forming NOAA was to create a civil center of strength for expanding effective and rational use of ocean resources, for monitoring and predicting conditions in the atmosphere, ocean, and space, and for exploring the feasibility and consequences of environmental modification.

Two of the most important components of NOAA are the National Weather Service (NWS), formerly the ESSA Weather Bureau, and the National Environmental Satellite Service (NESS), formerly the National Environmental Satellite Center of ESSA. The NWS reports the weather of the United States and its possessions, provides weather forecasts to the general public, issues warnings against tornadoes, hurricanes, floods, tsunamis (seismic-generated waves), and other atmospheric and hydrologic hazards, and provides a broad array of special services to aeronautical, *maritime*, astronautical, agricultural, and other weather- and ocean-sensitive activities. These services are supported by a national network of observation and forecasting stations, communications links, computers, aircraft, and satellite systems.

The National Environmental Satellite Service (NESS) operates the national weather and environmental satellite systems. It develops new techniques for the acquisition of environmental data and the application of such data to atmospheric, oceanographic, solar, and other geophysical problems.

The NWS provides reporting, forecasting, and warning services designed to promote navigation safety among both private and commercial seamen. Weather maps, warning display signals, Marine Weather Services Charts, weather broadcasts by commercial and marine radio stations, are just some of the important means used to provide these services.

Weather forecasts for boating areas in the United States and Puerto Rico are issued every six hours by the NWS. Each forecast covers a specific coastal area—for example, "Cape Hatteras N. C. to Savannah, Ga." If strong winds or sea conditions hazardous to small-boat operations are expected, the forecasts include a statement as to the type of warning issued and the areas where warning signals are displayed. The various types of warnings issued and the displays for each were described earlier in this chapter.

Similar forecasts and warnings are issued for numerous inland lakes, dams, reservoirs, and river waterways throughout the country. Daily information that indicates expected stream flow, river gauge heights, and flood warnings is also issued by the NWS, as necessary.

Special announcements are also provided by the NWS via radio and television broadcasts and the press whenever a tropical storm or hurricane becomes a potential threat to a coastal area. This is termed a *hurricane watch* announcement. Remember, it is *not* a hurricane warning. The hurricane watch announcement means that a hurricane is near enough so that everyone in the area covered by the watch should pay attention, listen for subsequent advisories, and be ready to take precautionary action *in case* hurricane warnings are issued in the near future. No visual display is provided for hurricane watch announcements.

Latest forecasts of all types issued are available over AM and FM radio, television, and marine radiotelephone broadcasts. Radio stations in cities located along principal rivers include stream flow and river data in their weather broadcasts. Whenever storm and flood warnings are issued by the NWS, all stations make frequent broadcasts of these conditions as a service to small-boat operators, the general public, and other interests.

The broadcast schedules of radio stations, NWS office telephone numbers, the locations of warning display stations, and all sorts of good and pertinent information for boatmen is reprinted on the Marine Weather Services Charts. These charts are issued periodically for the following areas:

MSC–1   Eastport, Me. to Montauk Point, N. Y.
MSC–2   Montauk Point, N. Y. to Manasquan, N. J.
MSC–3   Manasquan, N. J. to Cape Hatteras, N. C.
MSC–4   Cape Hatteras, N. C. to Savannah, Ga.
MSC–5   Savannah, Ga. to Apalachicola, Fla.
MSC–6   Apalachicola, Fla. to Morgan City, La.
MSC–7   Morgan City, La. to Brownsville, Tex.
MSC–8   Mexican Border to Point Conception, Calif.
MSC–9   Point Conception, Calif. to Point St. George, Calif.
MSC–10 Point St. George, Calif. to Canadian Border
MSC–11 Great Lakes: Michigan and Superior
MSC–12 Great Lakes: Huron, Erie, and Ontario
MSC–13 Hawaiian Waters
MSC–14 Puerto Rico and Virgin Islands
MSC–15 Alaskan Waters

Copies of the Marine Weather Services Charts are available at local marinas, at marine chart dealers, or you can get them by ordering from: National Ocean Survey, Distribution Division (C-44), 6501 Lafayette Avenue, Riverdale, Maryland 20854, (Price 25¢). Remember that Small Craft Advisories cover a wide range of wind and sea conditions, and the "small craft" category includes boats of many sizes, displacements, shapes, and designs. The Small Craft Advisory should alert every mariner to determine immediately the reason for the advisory or the display by tuning his radio to the latest marine broadcast. Deciding the magnitude of the danger is the responsibility of the mariner and will be based on his experience level and the type and size of boat.

The NWS also provides boatmen with continuous broadcasts of the latest weather information from 66 different locations in the continental United States, Alaska, and Hawaii. These NOAA VHF-FM Radio Weather transmissions repeat taped messages every four to six minutes. The tapes are updated about every two to three hours, and modified as necessary to include the latest information. The messages include weather and radar summaries, wind observations, visibility, sea and lake conditions, detailed local and area forecasts, and other information tailored to the needs of professional and amateur boatmen. When severe weather warnings are required, the routine transmissions are interrupted and the broadcast is devoted to emergency warning operations.

The NOAA VHF-FM Radio Weather transmissions are broadcast 24 hours per day at frequencies of 162.55 and 162.40 MHz from NWS offices across the nation. MHz = MegaHertz; one Hertz = one cycle per second; MegaHertz = one million Hertz = one million cycles per second. As an added refinement, the NWS forecasters can turn on specially designed radio receivers by means of a tone signal. This signal is transmitted at 1050 Hertz for three to five seconds before announcements of hazardous weather conditions. The tone signal alerts individuals and news media to be ready for critically important weather messages. The emphasis of all NWS effort is on public safety.

The NOAA VHF-FM broadcasts can usually be received 20–60 miles from the transmitting antenna, depending on terrain and the quality of the receiver used. The frequencies of 162.55 and 162.40 MHz lie above the commercial FM frequencies (which end at 108 MHz). Receivers are available in a variety of types and prices. In selecting a suitable receiver, pay special attention to the manufacturer's rating of the receiver's *sensitivity*. A receiver with a sensitivity of one microvolt *or less* should pick up a broadcast at a distance of 40–60 miles.

A "listening guide" to be used when buying a receiver for the reception of NOAA VHF Radio Weather is contained in table 1–1. The data used in constructing table 1–1 are based on the following parameters: (1) Transmission over clear terrain, (2) Standard 300-watt transmitting station, (3) Transmitter antenna height 300 feet, (4) Receiver antenna height six feet, and (5) Narrow-band FM receiver (deviation 15 KHz). KHz = KiloHertz; KiloHertz = one thousand Hertz = one thousand cycles per second.

All boatmen and mariners should be familiar with and subscribe to the publication, *Notice to Mariners*. It is published weekly by the Defense Mapping Agency (DMA) Hydrographic Center (DMAHC) and is prepared jointly by the NOAA's National Ocean Survey (NOS) and the U. S. Coast Guard (USCG) of the De-

Table 1–1. A guide to be used when buying a radio receiver for the reception of NOAA VHF Radio Weather.

Distance of Receiver from Transmitter Site	Sensitivity of Receiver
60 miles	0.6 microvolts
50 miles	0.9 microvolts
40 miles	1.2 microvolts
30 miles	2.5 microvolts
20 miles	6.0 microvolts
10 miles	20.0 microvolts

Courtesy of NOAA National Weather Service

partment of Transportation. All issues contain a wealth of information of interest to mariners of all descriptions and levels of expertise. Important contributions are also made by the U. S. Army Corps of Engineers and by foreign hydrographic offices and cooperating observers of all nationalities. The publication is global in scope.

The contents of *Notice to Mariners* are presented in three sections. Section I contains: Catalog Corrections, Geographic Index, Chart Corrections, New Charts and Publications, Chartlets and Depth Tabulations, Sailing Directions and Coast Pilot Corrections. Section II contains: Light List Corrections, Radio Aids Corrections, Other Publication Corrections. Section III contains: Broadcast Warnings and Marine Information, Miscellaneous.

Issues of *Notice to Mariners* may be obtained from local offices of the DMA, NOS, or the USCG, or by writing to one of the following: Defense Mapping Agency, Hydrographic Center, Washington, D. C. 20390, Attention: Code NS; U. S. Coast Guard, Washington, D. C. 20590, Attention: GWAN–174; National Ocean Survey, Rockville, Maryland 20852.

Another excellent publication which is a must for all who sail upon the sea is the periodical, *Mariner's Weather Log*, published bi-monthly by NOAA's Environmental Data Service. Each issue is loaded with photographs, charts, diagrams, and other illustrations covering subjects of interest to boatmen. Among other things, each issue contains fascinating weather-related articles, hints to observers, tips concerning radio operations, special subjects, interesting letters to the editor, marine weather reviews of the North Atlantic and North Pacific, and monthly marine weather diaries. *Mariner's Weather Log* issues are available from: Environmental Data Service, Page Building 1 (D762), 2001 Wisconsin Avenue, Washington, D. C. 20235.

Weather and oceanographic information, forecast information, and warning information are also readily available on a worldwide scale. Ships and land stations, regardless of location, observe and record weather conditions simultaneously every six hours as a matter of routine, starting at midnight Greenwich time. Under certain conditions, such as storm, hurricane, or unusual weather phenomena, weather and sea observations and reports are made at more frequent intervals. These reports are transmitted by landline or radio to designated collection and control centers and from there are relayed farther in order to reach all users in the shortest possible time. These reports are then used in the preparation of weather charts, from which forecasts are made and advisories or warnings issued. The forecasts and warnings are transmitted on assigned frequencies and at specified times to any recipient who can receive the broadcast. In addition to forecasts and warnings, a general description of the weather map is also broadcast, along with selected surface weather reports from which individuals aboard ship may prepare their own weather maps, if they so desire. Advancements in radio facsimile techniques have made possible the broadcast of pictures (facsimiles) of current and predicted (prognostic) weather charts. Innumerable merchant and naval ships of many countries are equipped to copy this form of radio transmission, thereby saving many man-hours of analyzing weather charts.

Complete weather broadcasts and their schedules for all parts of the world are contained in three publications. The first, *Worldwide Marine Weather Broadcasts*, is now the principal source of information on marine weather broadcasts. It has replaced Publication H. O. 118, published for many years by the Naval Oceanographic Office. It is revised annually, and interim changes and corrections are made by means of the weekly *Notice to Mariners* issued by the DMAHC. Section 1 lists the details of all U. S. and selected foreign radio stations that transmit scheduled radiotelegraph and radiotelephone weather broadcasts. Also included in this section is the Radio WLO, Mobile, Ala., radioteleprinter broadcast. A listing of weather broadcasts by U. S. radio stations on the Great Lakes precedes the pages for the North Atlantic. Section 2 contains details of U. S. and foreign radiofacsimile weather chart transmissions that are useful to ships and craft equipped to receive them. Section 3 lists the VHF–FM continuous-voice weather broadcasts within the U. S. These broadcasts are of great value to ships and craft entering and departing port. *Worldwide Marine Weather Broadcasts* is for sale by: Superintendent of Documents, U. S. Government Printing Office, Washington, D. C. 20402, Price: $1.45, Stock Number 0317–00147.

The second publication is, *International Meteorological Codes and Worldwide Synoptic Broadcasts*. This publication has replaced Section 3 (Synoptic Broadcasts) and Section 5 (Codes) of the now obsolete H. O. Publication 118, Radio Weather Aids. Changes to this new publication are provided by the Director, Naval Oceanography & Meteorology. Copies may be obtained from: Director, Naval Oceanography & Meteorology National Space Technology Laboratories, Bay St. Louis, Miss. 39520.

The third publication is *Information for Shipping*, which contains details of all foreign weather broadcasts. Its identifier is: Publication Number 9, TP-4, Volume D, and it is available from: The World Meteorological Organization (WMO), Case Postale No. 7, CH 1211, Geneva 20, Switzerland.

THE HURRICANE- AND TYPHOON-WARNING SYSTEMS

Except for the tornado, the hurricane or typhoon is the most violent and the most destructive meteorological phenomenon experienced by man—appropriately termed "nature on a rampage." These beasts of nature pack a fantastic punch. But this is not surprising when one considers the tremendous amounts of energy involved. In a single day, the average hurricane or typhoon releases an amount of heat which, if converted to electrical energy, could provide a six months' supply of electricity for the entire United States. In a single day, even a small hurricane or typhoon will release about 20 billion tons of water—an energy equivalent of almost 500,000 atomic bombs (or almost six atomic bombs per second)!

All sorts of interesting and awesome energy-equivalent calculations have been made by scientists, and all point in the same direction—hurricanes and typhoons (and their effects) are to be avoided, if at all possible.

One might logically ask, "Since there are only about 5 to 8 hurricanes per year in the Atlantic, 3 to 7 hurricanes per year in the Eastern and Central Pacific, and 19 to 25 typhoons per year in the Western Pacific, why all the fuss about hurricanes and typhoons?" The answer is quite simple.

Even though these tropical cyclones are *relatively* rare occurrences, they exact a tremendous annual toll in human life, in property damage, and in money. Since the turn of the century, more than 20,000 Americans have lost their lives to hurricanes. And it is likely that the annual hurricane damage in the United States, alone, will average more than $100 million. Once in every 10 or 15 years, a *single* tropical cyclone will cause property damage in excess of $1 billion!

Killer hurricanes and typhoons can occur in almost any year, and it is impossible to predict well in advance, with accuracy, when these devastating phenomena will develop. In October 1962, a typhoon crossing southern Thailand claimed almost 5,000 lives. Hurricane *Flora*, in 1963, was one of the worst storms in the Caribbean and Western Atlantic since Columbus discovered the New World. An estimated 7,200 lives were lost in Haiti, Cuba, and the Dominican Republic. In 1967, a single tropical cyclone in the Indian Ocean was responsible for the loss of over 3,500 lives. A similar cyclone and its storm surge in 1970 claimed almost one million lives in Bangladesh. In September 1974, Hurricane *Fifi* skirted the coast of Honduras, resulting in an estimated 7,000 deaths. And in the two-month period of October–November 1974, seven different typhoons crossed the island of Luzon in the Philippines, claiming countless lives!

This is why there is so much fuss about these meteorological hazards and why the threat which they constitute is not to be minimized.

Like the old Navy adage, "different ships, different long splices," tropical cyclones in various parts of the world are known by different names. But they are all the same basic weather phenomenon:

Area	*Name*
Atlantic	Hurricane
Caribbean Sea	Hurricane
Gulf of Mexico	Hurricane
Eastern Pacific	Hurricane
Central Pacific	Hurricane
Mexico	Cordonazo
Haiti	Taino
Western Pacific	Typhoon
Philippines	Baguio or Baruio
Indian Ocean	Cyclone
Australia	Willy-Willy

The earliest mention of the type of storm we now call *hurricane* appears in the voyage logs of Christopher Columbus, but he used the same terminology to describe the severe storms of both summer and winter, which are really quite different weather phenomena. The winter storms have a cold core (or center), whereas hurricanes (or any tropical vortex) have a warm core. It is generally believed that Columbus first encountered a hurricane in the vicinity of Santo Domingo in October, 1495. Some sources mention a hurricane in about the same geographical area in 1494. And in 1502, Columbus wrote about another hurricane —this one also in the vicinity of Santo Domingo. For a wealth of historical information on these very early hurricanes, the reader is referred to a wonderful book titled simply, *Hurricane*, by Marjory Stoneman Douglas (New York; Rinehart and Co.), published in 1958.

Through the years, real progress in hurricane and typhoon forecasting has been a slow and gradual affair, beset with many difficulties. To begin with, these storms are born and spend most of their lives at sea. Until the advent of radio, it was next to impossible to obtain observational weather data and reports in oceanic and tropical areas. In their excellent book titled, *Atlantic Hurricanes*, published in 1960, Gordon E. Dunn (then Director, National Hurricane Center) and Banner I. Miller (Research Meteorologist, National Hurricane Center) report the following:

"Throughout the first decade of the present century, [hurricane] detection was a most difficult problem. When a message was received by cable indicating the presence of a hurricane, warnings or a hurricane alert were frequently issued for extensive portions of the United States coastline. On one occasion during the first years of the 20th century a message was received in Washington, D. C., indicating that a hurricane was approaching the Antilles. During the ensuing week, hurricane warnings were displayed at one time or another from Charleston, South Carolina, to Brownsville, Texas, only to have the hurricane eventually show up at Bermuda. As late as 1909, hurricane warnings flew from Mobile, Alabama, to Charleston, South Carolina, in connection with a hurricane which eventually affected only extreme southern Florida. Not until about 1920 did forecasters begin to make a serious attempt to try to actually forecast hurricane paths."[*]

Fortunately for all mariners (and coastal and island residents, as well), the highly suspect, shotgun-type hurricane and typhoon forecasts and warnings are a thing of the past. From very humble beginnings, the former U. S. Weather Bureau (now the NOAA National Weather Service) and the Department of Defense have evolved a highly effective surveillance system of the hurricane areas of the Atlantic, Caribbean, Gulf of Mexico, Eastern and Central Pacific ocean areas. The same type of highly effective surveillance is conducted by the U. S. Department of Defense in the typhoon areas of the western North Pacific and Indian Ocean areas. Routine and special weather reports from land stations, ships at sea, and aircraft; weather satellite reports; radar reports from land stations, ships at sea, and aircraft; and the specially instrumented weather reconnaissance

---

[*]G. E. Dunn and B. I. Miller, *Atlantic Hurricanes*, (Baton Rouge: Louisiana State University Press, 1960), p. 137.

aircraft of the NWS of NOAA and the Department of Defense (DOD) all combine to permit the pinpoint location, detection, and tracking of hurricanes and typhoons with phenomenal accuracy. As one would expect, international cooperation in this endeavor is excellent.

To provide timely and accurate information and warnings regarding hurricanes and typhoons, the North Atlantic and North Pacific oceans have been divided into geographical areas of responsibility as shown in figure 1-9. The stars show the locations of the Navy's Fleet Weather Centrals engaged in this work at Norfolk, Virginia, at Pearl Harbor, Hawaii, and on the island of Guam, Marianas Islands, in the western Pacific. Those major Hurricane Warning Offices (HWOs) of the National Weather Service shown by circles in figure 1-9 are located at San Juan, Puerto Rico; New Orleans, Louisiana; Washington, D. C.; Boston, Massachusetts; San Francisco, California; and Honolulu, Hawaii. The NWS National Hurricane Center is located at Miami, Florida. While the responsibilities of NOAA's National Weather Service are to the general public and other interests, those of the Department of Defense are to meet the diverse and unique requirements of the U. S. military establishment. However, extremely thorough and complete coordination in all facets of this work is effected amongst the civilian and military meteorological organizations. Details of the hurricane- and typhoon-warning systems will be discussed in the following sections.

The Hurricane-Warning System

The geographical areas of responsibility for tropical cyclone/hurricane forecasting and warning are assigned to the National Weather Service (NWS) Hurricane Warning Offices (HWOs) and Hurricane Centers (HCs) as follows:

Caribbean Sea, Gulf of Mexico, and Atlantic Ocean:
   HWO San Juan, P. R.: Caribbean Sea, islands, and ocean areas south of latitude 20° North and longitudes 70° West to 55° West (warning responsibility only);
   HWO New Orleans, La.: Gulf of Mexico and its coasts west of longitude 85° West and north of latitude 25° North (warning responsibility only);
   HWO Washington, D.C.: Coastal and ocean areas from latitude 35° North to 41° North and eastward to longitude 65° West (warning responsibility only);
   HWO Boston, Mass.: Coastal and ocean areas north of latitude 41° North and West of longitude 65° West (warning responsibility only);
   NHC Miami, Fla.: *Forecast responsibility* for all coastal and ocean areas. Warning responsibility for all areas in the Gulf of Mexico and Caribbean Sea not assigned to HWO New Orleans or HWO San Juan, and those areas in the Atlantic Ocean not assigned to HWO Boston or HWO Washington.
Eastern Pacific:
   The Eastern Pacific Hurricane Center (EPHC) at San Francisco has the *hurricane forecasting* and *warning responsibility* for the eastern Pacific Ocean east of longitude 140° West, and north of the equator.

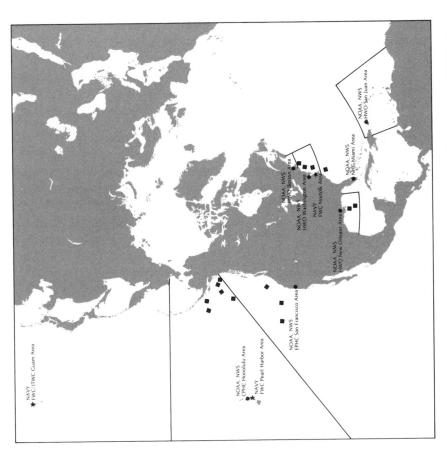

Figure 1–9. Geographical areas of responsibility of the major NOAA National Weather Service (NWS) (dots); Offices and Centers and Navy Fleet Weather Centrals (stars) for hurricanes and typhoons. The "squares" show the locations of NOAA's Experimental Data Buoys (EBs). Department of Defense.

Central Pacific:

The Central Pacific Hurricane Center (CPHC) at Honolulu has the *hurricane forecasting* and *warning responsibility* for the Central Pacific Ocean from longitude 140° West westward to the 180th meridian, and north of the equator.

Note from the preceding paragraphs that the NWS "centers" have both forecasting and warning responsibilities, while the "offices" have only warning responsibilities. If visualizing the latitudes and longitudes is a nuisance, refer to figure 1–9 which delineates the geographical areas of responsibility.

Within the Department of Defense, the Fleet Weather Central Norfolk has the forecast and warning responsibility for the entire Atlantic Ocean, Caribbean, and Gulf of Mexico areas for naval units. For both the Eastern and Central Pacific areas, the Fleet Weather Central Pearl Harbor is assigned with forecast and warning responsibilities for Navy ships.

The lists of hurricane names are contained in Appendix C.

In the Atlantic, tropical depressions are numbered as soon as their identity has been established; the first one being numbered "one." The NHC Miami assigns the numbers. In the Pacific, for the area east of longitude 140° West, the list of tropical depression numbers is maintained by the EPHC San Francisco. Numbering starts at the beginning of each calendar year. For the area west of longitude 140° West, the list of tropical depression numbers from 01 through 99 is maintained by the Fleet Weather Central/Joint Typhoon Warning Center (FWC/JTWC) Guam. Renumbering is at the end of sequence or, in all cases, at the beginning of each calendar year.

In the Atlantic and eastern Pacific, a separate set of names is used each calendar year, beginning with the first name in the set. In 10 years, after the 10 sets have been used, the same 10 sets are used again, in the Atlantic. In the Eastern Pacific, only four sets of names are used, as shown in Appendix C. In four years, after the four sets have been used, the same four sets are used again. Names beginning with the letters, Q, U, X, Y, and Z are not included because of the scarcity of suitable names beginning with these letters.

In the central North Pacific, when a tropical depression intensifies into a tropical storm or hurricane (between longitude 140° West and the 180th meridian), the CPHC Honolulu requests a name from the FWC/JTWC Guam. Then the depression number is discontinued and replaced by the appropriate name. For tropical cyclones originating east of longitude 140° West, the names are assigned by the EPHC San Francisco. Tropical cyclones that cross longitude 140° West from either west or east retain their original assigned name.

In the Atlantic Ocean, tropical storm and hurricane advisories for marine interests (Marine Advisories) are prepared by the NHC Miami at 0400, 1000, 1600, and 2200 Greenwich Mean Time. The Marine Advisories are edited by the Weather Service Forecast Office (WSFO) Washington, and then included in Part I (Warnings) of the weather broadcasts for high-seas shipping. Coastal radio stations transmit warnings and forecasts for offshore and coastal waters. In the Pacific Ocean, Marine Advisories are prepared by the EPHC San Francisco and the CPHC Honolulu at 0300, 0900, 1500, and 2100 Greenwich Mean Time.

These are edited and included in Part I (Warnings) of the weather broadcasts for high-seas shipping and are transmitted by coastal radio stations, as in the Atlantic. Warnings and forecasts for offshore and coastal waters are also transmitted.

Complete details of all these broadcasts are contained in the publication, *Worldwide Marine Weather Broadcasts*, mentioned earlier.

An extremely interesting source of information for weather and hurricane forecasters is the experimental environmental data buoys deployed in the Gulf of Mexico and off the coasts of the United States. These buoys gather environmental and engineering data needed for buoy test and evaluation; for improved buoy design; and for use in environmental monitoring, prediction, and research.

The experimental environmental data buoys (EBs) are deployed at the following locations:

EB–01	35.0° North	72.0° West
EB–03	56.0° North	148.0° West
EB–04	26.0° North	90.0° West
EB–15	32.3° North	75.3° West
EB–16	42.5° North	130.0° West
EB–17	52.0° North	156.0° West
EB–19	51.0° North	146.0° West
EB–20	41.0° North	138.0° West
EB–21	46.0° North	131.0° West
EB–33	59.3° North	140.0° West
EB–34	40.1° North	73.0° West
EB–35	55.3° North	157.0° West
EB–41	38.7° North	73.6° West
EB–61	28.5° North	90.9° West
EB–70	59.5° North	142.3° West

These buoys are 40-foot discus buoys and provide scheduled meteorological and oceanographic data for the analysts and forecasters. So, if you come across one while you're cruising, you'll know it isn't something from outer space which landed on the surface of the sea.

The hurricane-warning service of the United States has three principal functions: (1) the collection of the necessary observational data, (2) the preparation of timely and accurate forecasts and warnings, and (3) the rapid and efficient distribution of advices, warnings, and all other pertinent hurricane information to all civilian and military consumers.

Warnings and advisories are prepared and distributed, or broadcast, by the various centers, offices, and centrals as soon as a tropical storm or hurricane is discovered and for as long as it remains a hurricane or a threat to life, property, ships, and smaller craft. The messages contain the position of the storm/hurricane, its intensity, its direction and speed of movement, and a description of the areas of strong winds. Also included is a forecast of future movement and intensity. When the storm center is over the open sea, cautionary advices are given ships and small craft; and when it is approaching or likely to affect any land area, details on when and where it will be felt and data on tides, floods, and maximum wind and gusts are included. Although the normal advisory and warning times are 0400, 1000, 1600, and 2200 GMT in the Atlantic, and 0300, 0900, 1500, and 2100 GMT in the Pacific, occasionally a rapid development or the receipt of later reports demand the issuance of interim advisories. After hurricane warnings have been issued for a particular area, bulletins may be issued as often as once every hour, or as often as required.

Widespread and rapid distribution is given these warnings, advisories, and bulletins by all available means of communications, including the hurricane tele-type systems and electronic computer data-links, which make them immediately available to all meteorologists almost simultaneously. Each local weather office then gives the warnings intensified distribution by radio, telephone, television, newspaper, and other means in its area of responsibility. Arrangements are made with local radio stations for immediate broadcast, often from microphones located in weather offices. All these hurricane advices are also sent immediately from the forecast centers to the press associations, which give them the highest priority distribution.

The Typhoon-Warning System

Situated atop Nimitz Hill on the western side of the beautiful island of Guam in the western North Pacific (near latitude 13.5° North, longitude 145.0° East), the Navy's Fleet Weather Central Guam, with its built-in Joint (Navy and Air Force) Typhoon Warning Center (JTWC), is responsible for all tropical storm and typhoon advisories and warnings from the 180th meridian westward to the main-land of Asia (as shown in figure 1–9). This responsibility, covering a vast oceanic area, applies to civilian and military interests alike.

Similar to the Fleet Weather Central at Norfolk, Va., the FWC/JTWC Guam is another ultramodern, extremely well-equipped, and frightfully busy weather command. In addition to all the standard meteorological and oceanographic equipment and capabilities, the FWC/JTWC Guam also has the equipment neces-sary to receive, record, and transmit weather satellite data and information. It also has an electronic computer laboratory capable of communications of over 4,000 words per minute.

Specifically, this command's mission as assigned by the military Commander in Chief Pacific, is as follows: (1) provide all tropical cyclone and typhoon ad-visories and warnings west of 180° longitude, (2) determine typhoon reconnais-sance requirements and priorities, (3) conduct investigative and postanalysis pro-grams, including the preparation of annual typhoon summaries, and (4) conduct tropical cyclone and typhoon forecasting and detection research, as practicable.

The efficiency of any system, whether it be theoretical or operational, is directly proportional to the accuracy, timeliness, density, and frequency of the data fed into the system. Thus, the typhoon-warning system—or any system—can only be as good as the data supplied to the system. Great reliance is placed upon the vital data obtained by manned aircraft reconnaissance and weather satellite programs of the Department of Defense.

Whenever a tropical depression, a tropical storm, or a typhoon is in existence in the western North Pacific area, serially numbered warnings bearing an *emer-gency* precedence are broadcast from the FWC/JTWC Guam every six hours at

0000, 0600, 1200, and 1800 GMT. These warnings include the location, intensity, movement, and a description of the areas of strong winds (i.e., the radius from center of the 30-, 50-, and 100-knot winds). The warnings also include the predicted intensity, direction, and speed of movement, and locations (12-, 24-, 48-, and 72-hour predicted positions) of the storm/typhoon center.

In the western North Pacific, the following *Typhoon Conditions of Readiness* apply and are strictly adhered to:

Condition Four: Threat of destructive winds possible within 72 hours.
Condition Three: Destructive winds possible within 48 hours.
Condition Two: Destructive winds anticipated within 24 hours.
Condition One: Destructive winds anticipated within 12 hours.

Figure 1–10. Eight major tropical storms, hurricanes, and typhoons around the world as seen simultaneously by a weather satellite. NOAA (formerly ESSA) National Environmental Satellite Service (NESS) Photo.

The lists of typhoon names for the western North Pacific are contained in Appendix C. The names appearing in the four columns are assigned in rotation, alphabetically. When the last name ("Winnie") has been used, the sequence begins again with "Alice." Occasionally, when a killer typhoon has occurred (such as Typhoon Karen, which struck the island of Guam in November, 1962), that particular name will be retired from the active list. Note that the name "Karen" does not appear in any of the four columns.

Civilian and military personnel alike in positions of authority refer to the FWC/JTWC Guam as the western Pacific's "guardian angel." No more need be said.

Some Rules for Safe Boating

Before setting out:

1. Check local weather and sea conditions.
2. Obtain the latest weather forecast for your area from radio broadcasts.
3. When warnings are in effect, don't go out unless you are completely confident that your boat or craft can be navigated safely under the forecast conditions of wind and sea.
4. Be cautious when you see warning displays at U. S. Coast Guard stations, yacht clubs, marinas, and at other coastal points.
5. Check your maps, charts, and radio gear.
6. Check fuel (if appropriate), compass, electrical system, and life preservers.

Underway:

1. Keep a weather eye out for:
   —the approach of dark, threatening clouds, which may foretell a squall or thunderstorm.
   —any steady increase in the wind or sea.
   —any increase in the wind velocity opposite in direction to a strong tidal current. A dangerous rip-tide condition may form steep waves capable of broaching a boat or small craft.
2. Heavy static on your AM radio may be an indication of nearby thunderstorm activity.
3. Check radio weather broadcasts for latest forecasts and warnings.
4. If a thunderstorm catches you underway:
   —stay below deck, if possible.
   —keep away from metal objects that are not grounded to the boat's protection system.
   —don't touch more than one grounded object at the same time (you may become a "shortcut'" for electrical surges passing through the protection system).
5. If a tropical storm or hurricane catches you underway, review the section, Rules for Precaution or Disengagement, in chapter 8, "Weather Disturbances and Storms," of this book.

2
general flow of air about the planet

"A windy March and a rainy April make May beautiful."

—E. Leigh 1657

The ancient Greeks had a word for almost everything, and the science of weather was no exception. The word for the earth's envelope of air, the *atmosphere*, derives from the Greek words *atmos*, which means vapor, and *sphaira*, which means sphere. To appreciate more fully the intriguing—and sometimes frustrating —large-scale and small-scale movement of air parcels around our planet, we should know a little about the atmosphere's composition and structure.

COMPOSITION OF THE ATMOSPHERE

Air, the material of which our atmosphere consists, is a mixture of many different gases, as shown in figure 2–1. The chemical composition of dry air is remarkably constant everywhere over the earth's surface and up to a height of approximately 65 miles. A sample of dry air contains about 78 percent (by volume) nitrogen, 21 percent oxygen, almost 1 percent argon, about 0.03 percent carbon dioxide and minute amounts of other gases listed in figure 2–1.

The air also contains a variable amount of water vapor, most of which is concentrated below 35,000 feet. The maximum amount of water vapor which the air can hold depends entirely upon the temperature of the air. The higher the temperature, the more water vapor the air can hold. At temperatures below

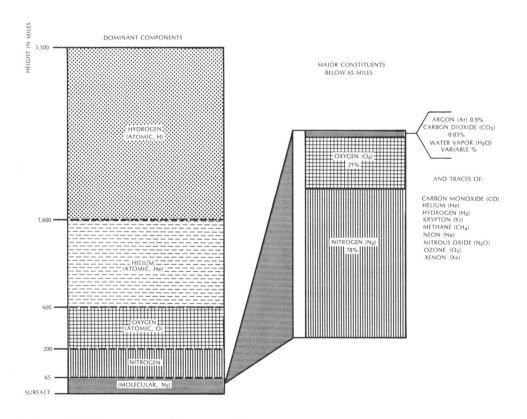

Figure 2–1. Composition of the atmosphere.

freezing, the amount of water vapor which can be held is very small. At high temperatures, it can amount to as much as 4 percent (by weight) of the air and water-vapor mixture. Water vapor is constantly being added to the atmosphere by evaporation from the earth's surface, particularly from the oceans, lakes, and rivers, and it is constantly being removed from the atmosphere by condensation resulting in precipitation in various forms, but mainly as rain and snow.

Air contains variable amounts of impurities such as dust, salt particles, smoke, other chemicals from sea spray, carbon monoxide, and micro-organisms. Dust, salt, soot particles, and the like are important because of their effect on visibility and especially because of the part they play in the condensation of water vapor resulting in precipitation, in one form or another. If the air were absolutely pure, there would be little condensation. These minute, impure particles act as nuclei for the condensation of water vapor. If the air were perfectly quiet, the heavier particles and gases would settle close to the earth and the lightest would be found the farthest out from the earth's surface. But the constant motion of the air near the surface mixes the gases, so that almost the same proportions exist throughout the atmosphere from the earth's surface upward to about 65 miles.

At a height of about 200 miles, the transition from a predominantly molecular atmosphere to an atomic one occurs, as atomic oxygen (O) displaces molecular nitrogen (N_2) as the dominant component. At roughly 600 miles, atomic helium (He) becomes the major component, and the region beyond 1,600 miles is dominated by atomic hydrogen (H). But in this space era, this is no longer "way out."

STRUCTURE OF THE ATMOSPHERE

All of us, mariner and landlubber alike, live most of our lives at the bottom of a virtual ocean of air which differs in one major way from an ocean of water; water is nearly incompressible. A cubic foot of water on the ocean bottom weighs about the same as a cubic foot of water near the surface. But the air of the atmospheric ocean *is* highly compressible. A cubic foot of air at the earth's surface weighs billions of times more than a cubic foot of air at the outer edge of the atmosphere. In fact, the atmosphere thins out so rapidly as one leaves the earth that when you are only 3½ miles above the earth, over half the atmosphere, by weight, lies below you. It is primarily in this 3½-mile envelope of heavy air that the weather changes are born.

For convenience, the atmosphere is divided into several layers, or *spheres*, according to their temperature (thermal) characteristics, as shown in figure 2–2. The turbulent layer near the earth in which nearly all weather is embedded is called the *troposphere*, which varies in thickness from about 5 miles over the poles to about 11 miles over the equator. In this layer, the temperature generally decreases with height. Above this layer, in the more stable *stratosphere*, the temperature generally increases with height, creating a sort of lid, or cap, which restrains the turbulence of the troposphere. Just above the stratosphere is the *mesosphere*, in which temperature again decreases with height. And finally, the *thermosphere*, the second region in which temperature increases with height, extends upward from the top of the *mesosphere*.

Between each thermal region there is a relatively sharp boundary zone identi-

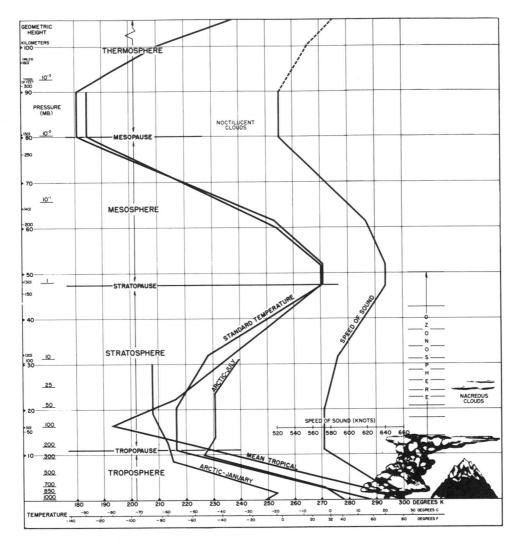

Figure 2–2. Structure of the atmosphere.

fied by the suffix, *pause*, added to the name of the layer below. Thus, the *tropo-pause* (at heights of 5 to 11 miles), the *stratopause* (at a height near 30 miles), and the *mesopause* (at a height near 50 miles) are the upper boundaries of the forenamed layers. This structural description of our atmosphere in terms of its thermal characteristics is just one useful approach. Other terms, found in other weather books, are based on such characteristics as composition, electron density, magnetic control, and so forth. For our purpose, the thermal approach is better.

To the right of the four-temperature profile (variation of temperature with height) curves in figure 2–2 is a profile of the speed of sound in air, a value which varies directly as the square root of the kinetic (resulting from molecular colli-sions) temperature. This variable has a great effect on the lift, drag, and control response of aerospace vehicles, especially at supersonic speeds. For this reason, the *Mach number*, generally defined as the ratio of the speed of the vehicle to the local speed of sound, is widely used to describe vehicle speeds in the supersonic range.

MOTIONS AND SEASONS OF THE EARTH

For our purposes, the shape of the earth can be considered a sphere. Actually, the earth is approximately an oblate spheroid (it is flattened at the poles and bulges at the equator). As a result, the earth's equatorial diameter (6,888.11 nautical miles) is larger than the polar diameter (6,864.92 nautical miles) by 23.19 nautical miles, and the meridians (of longitude) are slightly elliptical rather than circular as shown in Figure 2–3.

Of the earth's several motions of rotation, revolution, and translation in space, only two are important from the standpoint of weather: (1) the earth's rotation on its axis once every 24 hours results in day and night and generates the wind regimes of earth with their characteristic weather patterns, and (2) the earth's annual revolution around the sun in approximately 365.25 days results in the earth's seasons and their characteristic weather.

In the earth's (and every planet's) orbit around the sun, that point nearest the sun is termed the *perihelion*. The point farthest from the sun is termed the *aphelion*. And the line joining the perihelion and aphelion is called the *line of apsides*. As shown in figure 2–4, the earth is at perihelion in early January and at aphelion six months later in July.

The primary reason for our seasons is the 23.5° tilt of the earth's axis of rotation to the plane of its orbit around the sun. About 21 June, 10 or 11 days before reaching aphelion, the northern part of the earth's axis is tilted toward the sun, as shown in figures 2–4 and 2–5. The northern hemisphere polar regions are having continuous sunlight. The northern hemisphere is enjoying its summer with warm, long days and short nights. The southern hemisphere is having its winter with short days and cold, long nights. This is the *summer solstice*. About 23 Sep-

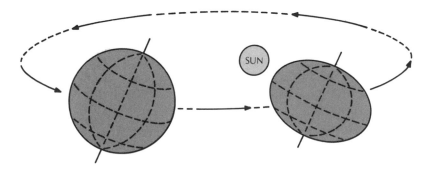

Figure 2–3. The earth as a sphere and as an oblate spheroid (exaggerated). On an oblate spheroid, the meridians (of longitude) are slightly elliptical rather than circular.

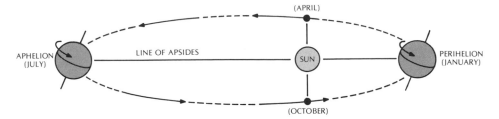

Figure 2–4. Orbit of the earth around the sun; perihelion, aphelion, and line of apsides. Note that the earth is at a greater distance from the sun during the summer months than during the winter months.

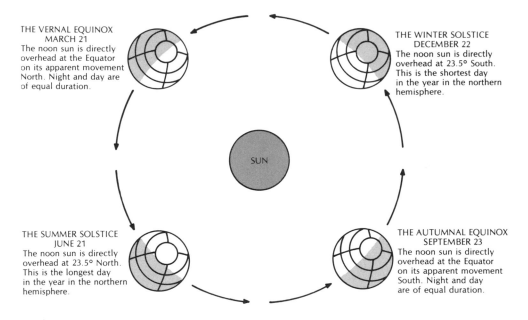

Figure 2–5. The seasons, solstices, and equinoxes of the northern hemisphere.

tember (three months later), the earth has moved one-fourth of the distance around the sun, but its axis of rotation still points in about the same direction in space. Both hemispheres receive equal amounts of sunshine, and the days and nights are of the same duration all over the world. The sun begins to set at the north pole, and rise at the south pole. The northern hemisphere is having its autumn, and the southern hemisphere its spring. This is called the *autumnal equinox*.

About 22 December (three months later), the southern hemisphere is tilted toward the sun. Conditions are now the reverse of those six months earlier. Now the northern hemisphere is struggling through winter, and the southern hemisphere is having its summer. This is called the *winter solstice*. About 21 March (again three months later), when the sun again shines equally on both hemispheres, the northern hemisphere is having spring, and the southern hemisphere is enjoying autumn. This is called the *vernal equinox*.

For those interested, the word "equinox" means "equal nights" and is used because it occurs at the time when days and nights are of about equal duration all over the world. The word "solstice" means "sun stands still" and is used because the sun stops its apparent northward and/or southward movement and "stands still," briefly, before it starts traveling in the opposite direction. It refers only to the movement in a north-south direction, and not to the daily apparent revolution around the earth.

At the time of the vernal equinox, the sun is directly over the equator, crossing from the southern hemisphere to the northern hemisphere. It rises due east and sets due west, and remains above the horizon about 12 hours. Following the vernal equinox, the sun climbs higher in the sky each day until the summer solstice. The sun then gradually retreats southward until it is again over the equator at the autumnal equinox.

It is perhaps surprising to some individuals that the sun is nearest the earth during the northern hemisphere winter. Consequently, it is *not* distance from the sun that is responsible for the difference in temperature during the different seasons. The reason centers on the altitude of the sun in the sky and the length of time the sun remains above the horizon. During the summer months, the sun's rays are more nearly vertical. Consequently, they are much more concentrated, as shown in figure 2–6. And since the sun is above the horizon more than half the time, heat is being added by absorption for a longer period than it is being lost by radiation. Astronomically, the seasons begin with the equinoxes and solstices. But meteorologically, they vary from place to place around the earth.

Everywhere between the parallels (of latitude) of about 23.5° N and about 23.5° S, the sun is directly overhead at some time during the year. Except at the extremes, this occurs twice: once as the sun appears to move northward, and the second time as it appears to move southward. The northern limit (23.5° N) is the *Tropic of Cancer*. The southern limit (23.5° S) is the *Tropic of Capricorn*. These names were derived from the constellations more than 2,000 years ago.

The latitude parallels at about 23.5° from the north and south poles mark the approximate limits of the circumpolar sun and are called the *polar circles*. The one in the northern hemisphere is termed the *Arctic Circle*. In the southern hemisphere it's the *Antarctic Circle*.

EFFECT OF TEMPERATURE AND
THE EARTH'S ROTATION ON WINDS

When air is not confined, changes in temperature result in changes in volume. Heated air becomes lighter (less dense) and expands. Cooled air becomes heavier (more dense) and contracts. If a large volume of air above the earth's surface is cooled, it contracts and becomes heavier causing a downdraft (called subsidence

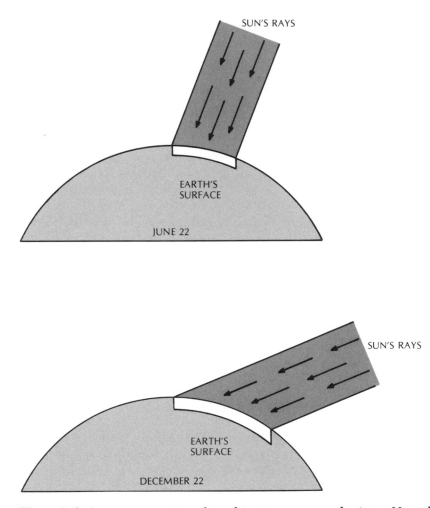

Figure 2–6. Area concentration of sunshine in summer and winter. Note the difference in the surface area covered by the same amount of sunshine; the area is much larger in winter.

by weathermen). The air from neighboring regions then moves horizontally to fill the void. This results in a greater mass of air over the region and the pressure is increased. In the reverse process, if a large volume of air near the earth's surface is heated, it expands and becomes lighter (less dense), causing an updraft (convection). This results in decreased pressure over the heated area.

Near the earth's surface, air tends to move from areas of high pressure toward areas of low pressure. In this way, a circulation is established. The air moves across the surface of the earth from areas of high pressure and low temperature toward areas of low pressure and high temperature, then vertically upward, then horizontally at high altitudes from areas of lower pressure toward higher pressure, where it travels vertically downward (subsides) to complete the circuit. The actual circulation of air over and around the earth is far more complicated than this because of, among other factors, the rotation of the earth and the continual changes in pressure and temperature.

If there were no heating or cooling, the temperature at any given altitude would remain everywhere the same. There would be no tendency for the air to move from one place to another. The air would lie at rest at the earth's surface. There would be no wind and no variation in "weather." However, as the result of the position and motion of the earth relative to the sun, and the physical processes involving radiation and absorption of energy, certain regions of the earth are always warmer than others. For similar reasons, the air over certain regions of the earth is seasonally warmer than over other regions. Also, local heating and cooling are continually taking place. The result of all this is that winds in some areas are relatively steady in both direction and speed. Others are seasonal. And both are continually modified by local conditions.

The heat that warms the earth's envelope of air is supplied primarily by the sun. As the radiant energy from the sun arrives at the earth, roughly 43 percent is reflected back into space by the atmosphere. About 17 percent is absorbed in the lower portions of the atmosphere, and the remaining approximately 40 percent reaches the surface of the earth. However, much of this is reradiated back into space. The earth's radiation is in long waves as compared to the short-wave radiation of the sun, since it emanates from a cooler body. Long-wave radiation is readily absorbed by the water vapor in the air and is primarily responsible for the warmth of the atmosphere near the earth's surface. In this way, the atmosphere acts like the glass of a greenhouse. It allows part of the sun's incoming short-wave radiation to reach the surface of the earth, but is heated by the terrestrial long-wave radiation passing outward. Over the entire earth and for long periods of time, the total outgoing energy must equal the total incoming energy (minus energy converted to another form and retained). Otherwise, the temperature of the earth, including its atmosphere, would steadily increase or decrease. This delicate balance is not required—nor does it exist—over short periods of time or in local areas.

As we saw in figure 2–6, more heat energy per unit area is received at the earth's surface when the sun's rays are closer to being perpendicular to the earth's surface. Measurements show that in the tropics more heat per unit area is received than is radiated away. In the polar regions, the opposite is true. If there were no process to transfer heat from the tropics to the polar regions, the tropics would be much warmer than they are, and the polar regions would be much colder. The process that accomplishes the necessary transfer of heat is called the *general circulation* of the earths' atmosphere.

The old adage, "If wishes were horses, beggars would ride," notwithstanding, *if* the earth had a uniform surface, did not rotate on its axis (but received equal amounts of sunshine at all points on the equator), and did not revolve around the sun (with its axis tilted at 23.5°), a rather simple circulation would result. This is shown in figure 2–7. However, this is not the case. The surface of the earth is far from uniform. Some of it is land. Much of it is covered by water. Some by ice. There are mountains and valleys. There are forests, deserts, and irregular coastlines. The earth rotates on its axis once in approximately 24 hours, so that the portion heated by the sun continually changes. The axis of rotation is

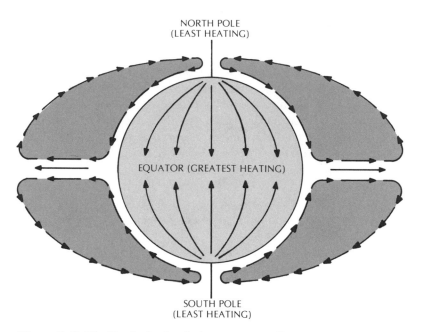

NORTH POLE
(LEAST HEATING)

EQUATOR (GREATEST HEATING)

SOUTH POLE
(LEAST HEATING)

Figure 2–7. Idealized air circulation over a uniform, nonrotating, nonrevolving earth.

tilted so that as the earth travels along its orbit around the sun, seasonal changes occur which result in variations in the heat balance. These factors, coupled with several others, result in complicated and constantly changing, large-scale movements of air over and above the surface of the earth.

The Horizontal Deflecting Force or Coriolis Force

At the equator, the distance around the earth is roughly 21,710 nautical miles. Since the earth rotates once in 24 hours, every point on the equator moves eastward at a speed of about 868 knots. With increasing distance (or latitude) from the equator, both north and south, each point on the earth's surface moves eastward more slowly. The distance it has to travel in 24 hours is less. At latitude 60°, for example, the earth spins eastward at about 434 knots. At the north and south poles, there is no eastward movement whatever because the poles are at the ends of the rotational axis.

In 1915, A. H. Compton devised a simple laboratory method to demonstrate the rotation of the earth. The apparatus consisted of a circular tube filled with water. The tube was placed in a vertical east-west plane and then rotated 180° about a horizontal east-west line. After the rotation of the tube, the water in the lower part had a tendency to flow eastward, because it had been brought down from a place of higher eastward speed to one of lower eastward speed. The water in the upper part of the tube had a tendency to flow westward for the opposite reason. The result was a very slow circulation of the water in the tube—westward at the top and eastward at the bottom.

The deflection of the wind (in both the northern and southern hemispheres) caused by the earth's rotation is known as the *Coriolis force* or the *horizontal deflecting force*. Referring to figure 2–8, let's assume that air is moving from point "X" northward to point "Y." As the air arrives at the latitude of point "Y," its distance from the rotational axis of the earth is considerably less than was its original distance. Since the moment of momentum of the mass of air remains constant (neglecting any forces acting on it in an east-west direction), it is moving eastward faster than the surface at latitude "Y." As a result, the mass of air moves northeastward, *not* northward, with respect to the surface of the earth. To an observer on earth, the air appears to have been deflected to the *right* in the *northern* hemisphere.

Now, let's assume that a mass of air is moving from the point "Y" southward toward the point "X." As the air moves southward, it loses eastward velocity because the point "X" moves eastward faster than the point "Y." Consequently, with respect to the surface of the earth, the mass of air moves southwestward, *not* southward. To an observer on earth, the air again appears to have been deflected to the *right* in the *northern* hemisphere.

For the same reasons, winds appear to be deflected to the *left* in the *southern* hemisphere.

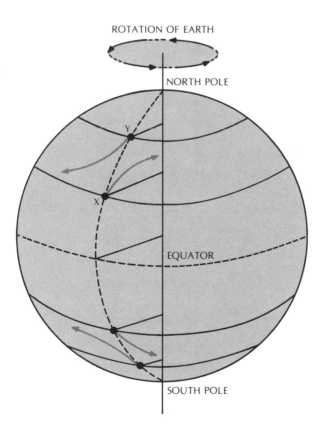

Figure 2–8. The Coriolis force. To an observer on earth, winds are deflected to the *right* in the *northern* hemisphere, and to the *left* in the *southern* hemisphere.

The general circulation of the earth's atmosphere in trade winds, anti-trade winds, and so forth, as we shall see in the next section, conforms to the foregoing requirements of the theory of the earth's rotation. *And the same is true of the larger ocean currents.*

When the density of the lower layers of the atmosphere is abnormally low because of high temperature relative to surroundings and/or large water vapor content, a condition of instability is generated. This condition is broken up sooner or later by an updraft of the heated layers from some limited area, accompanied by air flowing from all sides toward the region of the ascending currents. Each of the inflowing currents is deflected to the right (in the northern hemisphere) of the center of the updraft. The result is that a vortex is developed in which the motion is *counterclockwise* in the *northern* hemisphere (figure 2–9), and *clockwise* in the *southern* hemisphere.

In brief summary, the Coriolis force is strongest in the polar regions and zero at the equator. At intermediate latitudes, it varies directly as the sine of the latitude.

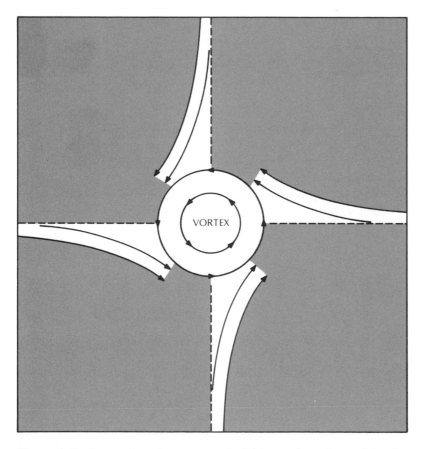

Figure 2–9. Generation of a counterclockwise cyclone (vortex) in the northern hemisphere.

GENERAL CIRCULATION OF AIR AROUND THE EARTH

Twenty-five years ago, the late Professor Victor P. Starr of the Massachusetts Institute of Technology wrote: "Since time immemorial man has inescapably observed the atmosphere in which he lives and has his being. It would therefore seem reasonable to expect that at the present date the science of meteorology should be one of the most advanced fields of human endeavor. Yet, if a distinction is made between the mere collection of descriptive facts of observation on the one hand and interpretive work that aims to give a rational intellectual understanding of phenomena on the other, it must be confessed that our knowledge concerning the large-scale motions of the atmosphere is restricted mostly to the former category of information . . . ."

That statement produced no counterarguments or debate throughout the years on the part of atmospheric scientists. However, we now have real reason for hope. The advent of fifth-generation (and later) high-speed computers, more complete global observations, meteorological rockets, advanced computer modeling, weather satellites, expanded international cooperation, and the World Weather Program should afford mankind a detailed knowledge and understanding regarding the general circulation of the earth's atmosphere in the foreseeable future.

By taking the statistical and climatological averages covering many, many years, a rough idea of the general circulation of the earth's atmosphere can be obtained. A simplified diagram of the general pattern is shown in figure 2–10.

The Doldrums—Near the equator, a "belt" of low pressure occupies a position about midway between the high-pressure belts found near latitude 30° to 35° in both hemispheres. Except for slight diurnal changes (occurring every day—four times), the atmospheric pressure along the equatorial trough is almost uniform. The wind is usually very light. And the light breezes that do blow are variable in direction. Hot and sultry days are a common occurrence. The sky is often overcast, and heavy showers and thundershowers are quite frequent. This equatorial low-pressure trough is relatively narrow, the eastern section being wider than the western section in both the Atlantic and Pacific oceans. Both the position and the extent of the trough vary with the seasons. During February and March, it lies just to the north of the equator and is quite narrow (a few miles in width). In July and August, the trough (or "belt") is centered at about latitude 7°–9° north, and sometimes penetrates as far as 20° N in the Western Pacific. At this time of year, it may be as much as a few hundred miles in width.

The Trade Winds—These winds blow outward from the zones or "belts" of high pressure near 30°N and 30°S toward the equatorial trough of low pressure. Because of the rotation of the earth, the moving air parcels are deflected toward the west (to the right in the northern hemisphere and to the left in the southern hemisphere). As a result, the trade winds in the northern hemisphere blow from the northeast and are called the *northeast trades*. In the southern hemisphere, they blow from the southeast, and are called the *southeast trades*. Over the eastern parts of both the Atlantic and Pacific oceans, these winds reach a much greater distance from the equator. Also, their direction is more N'erly and NNE'erly (in the northern hemisphere) and more S"erly and SSE'erly (in the southern hemisphere) than in the western parts of each ocean.

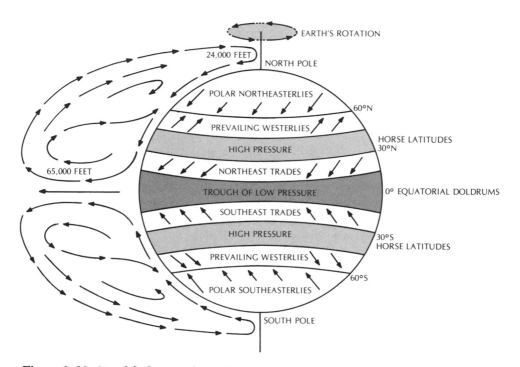

Figure 2–10. Simplified general circulation of air over the rotating earth. Compare this with figure 2–7.

The trade winds are perhaps the most constant winds on earth. But like many things in life, their constancy is often exaggerated. Occasionally they weaken or change direction, and there are regions where their normal pattern is disrupted. For example, around the island groups in the South Pacific, the southeast trades are almost nonexistent in January and February. Their greatest constancy occurs in the South Atlantic and in the South Indian Ocean. And everywhere they blow stronger in the winter months than in the summer. During July and August, when the equatorial trough of low pressure moves some distance north of the equator, the southeast trades blow across the equator, into the northern hemisphere, where the earth's rotation deflects them to the *right*, and they become southerly and southwesterly winds. The *southwest monsoons* of the Central American and African coasts originate partly in the southern hemisphere's diverted southeast trades. Cyclonic storms usually do not move into the regions of the trade winds. *But*, hurricanes and typhoons frequently form in these areas!

The Horse Latitudes—On the poleward side of each trade-wind belt, and corresponding roughly with the zone of high pressure in each hemisphere, is another region with weak pressure gradients and associated light and variable winds. These regions are called the *horse latitudes*. Unlike the weather in the doldrums, the weather is usually clear and fresh. Periods of "stagnation" are less persistent, more of an intermittent nature. The difference results primarily from the fact that the rising currents of warm air in the equatorial trough of low pressure carry aloft large amounts of moisture which condense as the rising air expands and cools at higher levels. In the horse latitudes, on the other hand, the air is descending (subsiding) and becomes less humid as it is warmed at lower altitudes.

The Prevailing Westerlies—On the poleward side of the high-pressure zone or belt in each hemisphere, the atmospheric pressure again decreases. The air currents moving toward the poles are deflected by the earth's rotation toward the east (to the right in the northern hemisphere and to the left in the southern hemisphere). They become southwesterly winds in the northern hemisphere and northwesterly winds in the southern hemisphere. These two wind systems are called the *prevailing westerlies* of the temperate zone in each hemisphere.

This relatively simple pattern is distorted considerably in the northern hemisphere because of the presence of large land masses. For example, between latitudes 40°N and 50°N in the North Atlantic, winds blow from some direction between northwest and south about 75 percent of the time, less persistent in summer than in winter. Also, they are stronger in winter, averaging about 25–30 knots as compared to 10–18 knots in summer. In the southern hemisphere, the westerlies blow throughout the year with a steadiness almost as great as that of the trade winds. Wind speed, although somewhat variable, is usually between 18–30 knots. These winds occur between latitudes 40°S and 55°S and are called the *roaring forties*. The higher speed and greater persistence of the westerlies in the southern hemisphere are caused by the difference in the atmospheric pressure pattern—and its variations—from that of the northern hemisphere. There is comparatively little land mass in the southern hemisphere, and the average annual pressure decreases much more rapidly on the poleward side of the high-pressure belt. There are fewer "irregularities" caused by the presence of large land masses (continents), as is the case in the northern hemisphere.

Winds in the Polar Regions—Temperatures are low near the geographical poles of our planet. Because of this, pressure tends to remain higher at the poles than in the surrounding areas. As a result, the winds blow outward from the poles and are deflected to the west by the rotation of the earth. They become *northeasterlies* in the *Arctic*, and *southeasterlies* in the *Antarctic*. They meet the prevailing westerlies near latitudes 60°N and 60°S. In the Arctic, the general circulation is modified considerably by the surrounding land masses. Winds over the Arctic Ocean are somewhat variable, and strong surface winds are not encountered very often. In the antarctic region, a high, central land mass is surrounded by water. This is a condition which augments, rather than diminishes, the general circulation in the area. The high pressure is stronger than in the Arctic, and has great persistence near the south pole. The upper-level air descends (subsides) over the high, south polar continent, where it becomes intensely cold. As it travels outward (and downward) toward the sea, it is deflected toward the west (to the left) by the earth's rotation. The winds remain strong throughout the year. They frequently reach hurricane and typhoon speed (greater than 64 knots or 74 mph). Sometimes they attain speeds as high as 200 knots (230 mph)!

With the exception of hurricane, typhoon, and tornado winds, these are the strongest surface winds encountered anywhere in the world.

Variations in the General Circulation—Various conditions modify the general circulation of the earth's atmosphere. Some are more understandable than others. All are complicated. Each has its own idiosyncrasies.

The Semipermanent Highs: The high pressure in the horse latitudes is really not distributed uniformly in zones or belts all around the world. In actuality, it is accentuated at various points around the globe, and appears as high-pressure cells (or anticyclones) on weather maps and charts. However, these high-pressure cells, or anticyclones, remain at about the same places with great persistence.

The Semipermanent Lows: These phenomena also occur in various places around the world. In the northern hemisphere, the prominent ones occur in the area to the west of Iceland and over the Aleutians in the winter. In the southern hemisphere, they are located in the Ross Sea and the Weddell Sea in the Antarctic. These areas are sometimes called by weathermen "the graveyards of lows," since many low-pressure systems move directly into these areas and are absorbed into the semipermanent lows which they reinforce. The low pressure in these areas is maintained largely by the migratory lows which come to a standstill in the region. It is also partly maintained by the sharp temperature difference between the polar regions and the warmer ocean areas.

The Monsoons: Another factor which causes variations in the general circulation is *land*. During summer months, a continent is warmer than its surrounding oceans. Also, it undergoes greater temperature *changes* than does the sea. As a result, lower pressures tend to prevail over the land. When a belt of high pressure encounters such a continent, its pattern is interrupted, or distorted. A belt of low pressure, on the other hand, is intensified. And the winds associated with belts of high pressure or low pressure are distorted accordingly. The opposite effect takes place in winter months. Belts of high pressure are intensified over land masses, and belts of low pressure are interrupted.

Perhaps one of the most striking examples of a wind system produced by the alternate heating and cooling of a land mass is the *monsoons* of the Indian Ocean and the China Sea. Referring to figure 2–11a, in the summer months, low pressure dominates the warm continent of Asia. High pressure dominates the adjacent sea. Between these two major wind systems, the winds blow in a nearly steady direction. The lower portion of the pattern extends to about latitude 10°S in the southern hemisphere. Here, the earth's rotation causes a deflection to the left, resulting in southeasterly winds. As the winds cross the equator, the deflection is to the right in the northern hemisphere, and the winds become southwesterly. In the winter months, figure 2–11b, the positions of the high- and low-pressure areas are reversed (high over land, and low over water) and, consequently, the air flow is reversed.

In the China Sea, the summer monsoon blows steadily from the southwest, from about May to September. These strong winds are accompanied by heavy squalls and thunderstorms. Rainfall is very much heavier than during the winter monsoon. As the season advances, rainfall and squalls become less frequent. In some places, the wind becomes light and unsteady. In other places, it continues reasonably steady. The winter monsoon blows from the northeast, from about October to April. It is quite a steady wind, frequently attaining a speed of 35

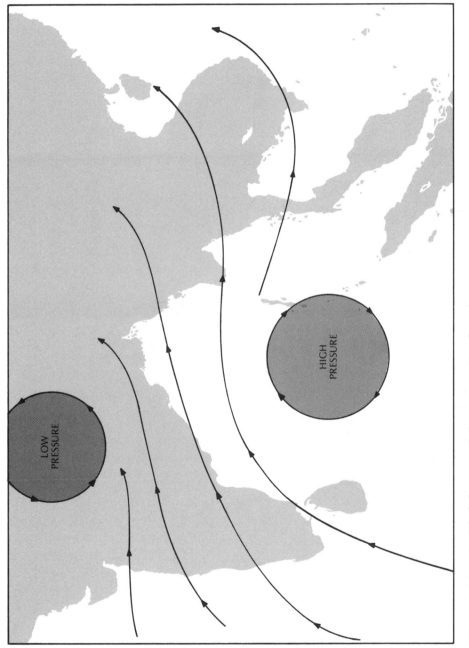

Figure 2–11a. Simplified diagram of the summer (southwest) monsoon.

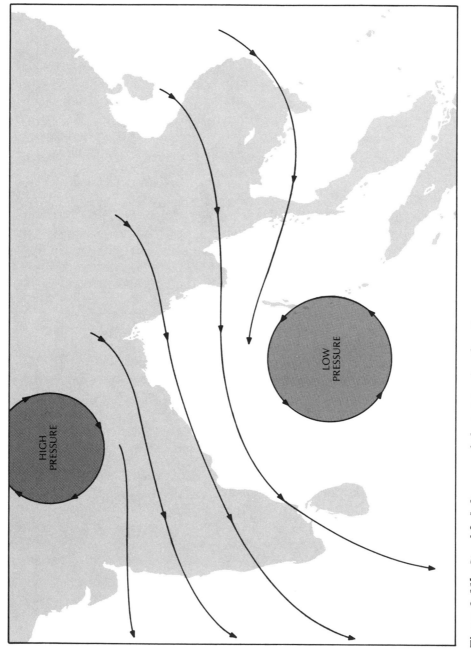

Figure 2–11b. Simplified diagram of the winter (northeast) monsoon.

knots (40 mph). Except for the windward slopes of mountainous or hilly areas, skies are generally clear to partly cloudy during the winter monsoon, and there is relatively little rain (except for the windward slopes of mountains). Also, some low cloudiness and/or light fog may occur along coastal areas.

SOME "WHY'S," "WHAT'S," AND "HOW'S" OF WEATHER

Air is the medium in which we live and "do our thing." Without it, life on this earth, as we know it, could not exist. We can live for many days without food, for a few days without water, but for only a few minutes without air. Air supports combustion and transmits sound. It is also a source of power, and when compressed, actuates various pneumatic systems and tools. The air comprising our atmosphere is indispensable to man.

Meteorology is defined as the science of the atmosphere and its phenomena. And these phenomena, collectively, we call the weather. *Weather* is the condition of the earth's atmosphere at any given time. There are a number of physical properties and conditions of the atmosphere that are observed and measured to describe the weather at a given time and place. The most important of these are: air temperature, atmospheric pressure, wind speed and direction, humidity (amount of water vapor, or moisture, in the air), amounts of cloudiness and sunshine, and precipitation.

Climate is the summation of weather conditions over a long series of years. It is not merely the average weather. Climate also includes the extremes and the variability of the weather elements. Since at any one place the weather is in constant flux, a long series of observations is needed in order to have reasonably accurate information concerning the average and most frequent conditions as well as the probable variations. One should remember that the normal, or average, value of a weather element is not necessarily its most probable value.

Water vapor is contributed to the air by evaporation from water surfaces (oceans, lakes, rivers, etc.), from the soil, and from living tissues. Like oxygen, it is one of the most important constituents of our atmosphere. But unlike the other gases, it is extremely variable in amount, ranging from minute proportions in the air over deserts to a little over 4 percent by volume in warm and humid air (such as in fog). Some water remains in the air as a gas (vapor) at all temperatures. And the higher the temperature, the more water vapor the air can "hold." The importance of atmospheric moisture to life and to the generation of "weather" cannot be overemphasized. Since water is a large and essential constituent of living organisms, the earth is habitable only because of the large amount of moisture at its surface and in its atmosphere.

Water changes its state readily in the atmosphere, from solid to liquid to gas, and also in the reverse order. The change from solid or liquid to gas is called *evaporation*, and *adds* water vapor to the air. The change from gas to liquid is called *condensation* and *removes* water vapor from the air. And the change from solid directly to gas is termed *sublimation*.

The amount of water vapor in the air may be expressed as its mass-per-unit volume, or its pressure-per-unit area. This is known as *absolute humidity*. The weight of water vapor present per unit weight of moist air is termed the *specific humidity*. The term we are most interested in, however, is the relative humidity.

Figure 2–12. The cloud patterns and general circulation of air around the earth, as seen by the NOAA-4 polar-orbiting weather satellite on 28-29 Dec. 1974. The upper picture shows the northern hemisphere; the lower picture shows the southern hemisphere. Note how the cloud circulation patterns are reversed in the southern hemisphere.

The *relative humidity* is the ratio (expressed in percent) of the actual quantity of water vapor to the saturation quantity at the same temperature, or the actual vapor pressure to the saturation vapor pressure. Expressed another way, the relative humidity is the ratio of the amount of water vapor which the air is "holding" to the amount which it *could* hold at that temperature, expressed as a percentage. Absolute and relative humidity change when the temperature or pressure changes, but specific humidity remains constant unless there is an actual loss or gain of moisture. The rate at which evaporation from a water surface occurs increases as the temperature and wind increase, and decreases as the relative humidity increases.

The condensation of water vapor in the upper levels of the atmosphere or near the earth's surface is the result of cooling. And this condensation takes many forms, such as clouds, rain, snow, fog, dew, and so forth. If the air is perfectly free of dust, it may be cooled far below its saturation point—or dew point—without condensation. The air is then said to be supersaturated. However, if smoke particles or salt spray from the oceans is present, rapid condensation occurs. With certain substances in the air, moisture begins to condense even before saturation is reached. Some of the ocean salts and some of the products of combustion have the quality of absorbing moisture from the air, and are said to be "hygroscopic." The presence of hygroscopic, or at least water-soluble, particles is essential to the condensation of moisture in the air. These particles are called "nuclei of condensation" or hygroscopic nuclei. And they are always present in the atmosphere in ample numbers.

Clouds are primarily the result of dynamic cooling produced by the expansion of air under reduced pressure as it rises vertically. The reduction of pressure at higher levels may sometimes produce a sufficient reduction of temperature to cause the formation of clouds. By far, however, the most important cause of clouds is the dynamic cooling resulting from the ascending air. Yet, some clouds are formed by the mixing of warmer and colder air.

Three conditions are necessary for the production of heavy rain: (1) cloud particles must be of different sizes (so that they will be moving at different speeds), (2) large quantities of water must be released under the influence of rising currents, and (3) the cloud must have large vertical extent (so that a falling drop will meet many smaller drops as it descends). As the drops descend, they grow mainly by combining with the cloud particles through which they fall. If any of the above conditions are absent, clouds may persist but there will be little or no precipitation.

Precipitation takes the form of rain when temperatures remain above freezing; snow, when condensation occurs below freezing, and the descending snowflakes remain frozen; hail, when the particles are subjected to alternate freezing and thawing; snow pellets, when soft snow is "compacted" into small white balls; sleet, when raindrops are frozen into icy "pellets." And glaze is the coating of ice formed by the freezing of rain as it comes in contact with a cold surface or object.

In addition to air and moisture, heat is an all-important ingredient for the production of "weather." Without atmospheric heating, mixing, and motion, there would be no weather, as we know it. And changes in weather result from temperature differences throughout the earth's atmosphere. Air, moisture, and heat—these are the basic ingredients of earth's weather.

3
basic cloud formations

"Look when the clouds are blowing
And all the winds are free:
In fury of their going
They fall upon the sea.
But though the blast is frantic,
And though the tempest raves,
The deep intense Atlantic
Is still beneath the waves."
                —Frederic William Henry Myers (1843–1901)

These few words express so magnificently the wonders and the excitement of the earth's clouds, weather, and oceans. Some of nature's most beautiful artistic displays are the various cloud formations. And understanding clouds—how, why, when, and where they form, what shape they take, and what the different types of clouds signify—is very important to mariners and anyone else involved with the weather and its changes. Especially to individuals who follow the sea, a knowledge of clouds is an indispensable tool.

Clouds assume an almost infinite variety of forms. They offer visual evidence of conditions existing in the atmosphere and also of changes that are taking place in the atmosphere. Consequently, clouds also afford an indication of future weather conditions, particularly if they are observed at various times and time intervals to note the changes in cloud structure or type that may be taking place. As one example, one of the first indications of the existence of an approaching mid-latitude storm or tropical hurricane or typhoon is the appearance of high-level cirrus clouds. As the storm or hurricane moves closer, high-level cirrostratus clouds are observed, followed by lower and thicker altostratus clouds. Finally, with the storm or hurricane at hand, low, dark clouds herald the arrival of high winds and torrential rains.

The portion of the earth's atmosphere which extends from the earth's surface to a height of about 24,000 feet above the poles, and to a height of about 65,000 feet above the equator, is called the *troposphere*. Of greatest interest to mariners are the clouds that form and go through their life cycles—some in a short time, others over longer intervals—in the troposphere. All clouds in this region are made up of either water droplets or ice crystals. And sometimes, a mixture of the two. It depends on the temperature. High clouds above the freezing level usually consist of ice crystals. Middle clouds consist primarily of water droplets. And low clouds are composed almost entirely of water droplets. Again, it is a matter of temperature. Clouds which extend from near the earth's surface to great heights in the troposphere consist of both water droplets (in the lower portion where temperatures are above freezing) and ice crystals (in the upper portion where temperatures are below freezing). More and more scientific knowledge on the fascinating subject of clouds is being obtained every day through the use of manned aircraft, high-altitude sounding balloons, meteorological rockets, controlled laboratory experiments, weather radars, and meteorological satellites.

CLASSIFICATION AND NAMES OF CLOUDS

Meteorologists around the world classify clouds on the basis of how the clouds are formed, the main characteristics, the peculiarities in shape and in internal structure, the special characteristics of arrangement and transparency, supplementary features, and so forth. For our purposes, we will concentrate on the aspect of how they are formed.

One principal type of cloud is called *cumulus*. This type is *formed by rising air* currents. It is usually dense and has a sharp, non-fibrous outline in the sky. Cumulus clouds appear as rising domes, or towers, and the upper parts often look like a cauliflower. The sun shining on these clouds makes them appear as a brilliant white mound with a relatively dark and horizontal base. A showery type of precipitation—when it occurs—is associated with cumulus-type (cumuliform) clouds.

Another principal type of cloud (or cloud genus) is called *stratus*. These clouds are in the form of a gray layer with a fairly uniform base. This type is *formed by the cooling of air below its saturation point with little or no vertical movement.* Whereas cumulus-type (*cumuliform*) clouds have extensive *vertical* development, stratus-type (*stratiform*) clouds have extensive *horizontal* development. The extensive horizontal layer, or layers, of stratus clouds need not necessarily be continuous. Stratus clouds do not always produce precipitation, but when they do, the precipitation is usually in the form of minute particles such as drizzle, snow grains, or ice crystals.

For scientific purposes, a necessarily complex cloud classification scheme was adopted and published by the World Meteorological Organization (WMO). For our operational and marine safety purposes, however, we can use a simple scheme and classify clouds according to the usual altitudes at which they are found: high, middle, and low. We should also be familiar with a few terms such as: *nimbus*, meaning rain-producing; *fracto*, meaning ragged, shredded, or torn; *alto*, used to indicate middle (and occasionally high) clouds; *cirri, cirro*, or *cirrus*, used to indicate high clouds; *virga*, meaning wisps or streaks of water or ice particles falling out of a cloud but evaporating before reaching the earth's surface as precipitation; and *cumulonimbus*, meaning thunderhead or thunderstorm. We can forget about uncinus, spissatus, lacunosis, perlucidus, and so forth.

Thus, our uncomplicated mariners' cloud classification scheme will be as follows:

High Clouds (above 18,000–20,000 feet)
Cirrus, cirrostratus, cirrocumulus, occasionally altostratus and the tops of cumulonimbus.

Middle Clouds (from about 7,000 feet up to 18,000–20,000 feet)
Altostratus, altocumulus, nimbostratus, and portions of cumulus and cumulonimbus.

Low Clouds (from near the ground up to about 7,000 feet)
Stratus, stratocumulus, most cumulus and cumulonimbus lower portions and bases, and frequently nimbostratus.

High Clouds (above 18,000–20,000 feet)

Description

Cirrus (Ci) clouds (figure 3–1) are detached wisps of hair-like (fibrous) clouds, formed of delicate filaments, patches, or narrow bands. Like the other high clouds, they are composed primarily of ice crystals. They are often arranged in bands which cross the sky like meridian lines and, because of the effect of perspective, converge to a point on the horizon.

Meaning

Cirrus clouds, that are scattered and are not increasing, have little weather meaning except to signify that any bad weather is at a great distance. Cirrus clouds in thick patches mean that showery weather is close by. These clouds are asso-

Figure 3–1. Cirrus clouds. Courtesy of R. K. Pilsbury.

ciated with, and formed from, the tops of thunderstorms. Cirrus clouds shaped like hooks or commas indicate that a warm weather front is approaching, and that a continuous-type rain will follow—especially if the cirrus is followed by cirrostratus. They also frequently indicate the presence and location of a jet stream.

Description

Cirrostratus (Cs) clouds (figure 3–3) are transparent, whitish clouds that look like fine veils or torn, wind-blown patches of gauze. They never obscure the sun to the extent that shadows are not cast by objects on the ground. Because they are composed of ice crystals, cirrostratus clouds form large halos—or luminous circles—around the sun and moon.

Meaning

Cirrostratus clouds, when in a continuous sheet and increasing, signify the approach of a warm weather front or an occluded weather front (discussed in chapter 7) with attendant rain or snow and stormy conditions. If these clouds are not increasing, and are not continuous, this means that the storm is passing to the south of you and no bad weather will occur at your location.

Description

Cirrocumulus (Cc) clouds (figure 3–4) are thin, white, grainy, and rippled patches, sheets, or layers showing very slight vertical development in the form

Figure 3–2. Jet stream cirrus clouds. Note that the condensation trail created by a jet aircraft flying at cloud level disappears in the drier air to the north. Coutesy of R. K. Pilsbury.

Figure 3–3. Cirrostratus clouds with cirrus, cirrocumulus, and lower altocumulus also present. Courtesy of R. K. Pilsbury.

Figure 3–4. Cirrocumulus clouds. Courtesy of R. K. Pilsbury.

of turrets and shallow towers. When these clouds are arranged uniformly in ripples, they form what seafaring men call a *mackerel sky*. These clouds are usually too thin to show shadows.

Meaning

Cirrocumulus clouds are quite rare and are of mixed significance. In some areas, these clouds foretell good weather; in others, bad weather. These clouds usually signify good weather along the west coast of the United States, in New England, and in the British Isles. They signify bad weather in most of southern Europe, particularly in Italy.

Middle Clouds (from about 7,000 to 20,000 feet)

Description

Altostratus (As) clouds (figure 3–5) are grayish layers of clouds. They are usually uniform in appearance and cover part, or all, of the sky. They are composed of water droplets or ice crystals, depending on the temperature. Neither the sun nor the moon shining through them form halos.

Meaning

Altostratus clouds are one of the most reliable weather indicators of all the clouds. They are indicative of warm air flowing up over colder air and impending rain or snow of the continuous type, especially if the overcast cloud layer progresses and thickens. These clouds are a good indication of a new storm development

Figure 3–5. Altostratus clouds. Courtesy of R. K. Pilsbury.

at sea. They frequently signal the formation of a stormy low-pressure area long before it is apparent from sea-level isobars or wind. What is important to sea-faring men is that these clouds are quite a reliable indication of approaching rain or snow, with associated poor visibility, large waves, and heavy swell.

Description

Altocumulus (Ac) clouds (figures 3–6 and 3–7) are most often seen as extensive "cloudlets" arranged in a regular pattern. Sometimes, when the sun is shining through these clouds, a corona can be observed. These clouds are usually composed of water droplets. Ice crystals are present only at very low temperatures.

Figure 3–6. Altocumulus clouds. Courtesy of R. K. Pilsbury.

Figure 3–7. Altocumulus clouds (Lenticularis). This is a classic example of a mountain-wave cloud taken near Taormina, Sicily. Courtesy of R. K. Pilsbury.

Meaning

Altocumulus clouds are significant primarily when they are followed by thicker high clouds or cumuliform low clouds. When they are in parallel bands, these clouds are in advance of a warm front with its associated steady rain or snow. When altocumulus clouds occur in the form of turrets rising from a common, flat base, they are usually the forerunner of heavy showers or thunderstorms.

Low Clouds (from near the ground to about 7,000 feet)

Description

Nimbostratus (Ns) clouds (figure 3–8) are the true rain and/or snow clouds, depending upon the temperature. These clouds are low, amorphous, dark, and usually quite uniform. They are thick enough to block out the sun, and they have a "wet look." When these clouds precipitate, the rain or snow is usually continuous. These clouds are often accompanied by low *scud* (fractostratus) clouds when the wind is strong.

Meaning

Nimbostratus clouds are of little help as a forecasting tool, since the bad weather is already at hand when these dark clouds with their associated heavy rain or snow are overhead. But if they are at some distance from you, and you have a report that they are coming your way, you know what to expect and can take

Figure 3–8. Nimbostratus clouds (with fractostratus below). Courtesy of R. K. Pilsbury.

the necessary precautions. Once these clouds have formed in your area or are heading your way, bad weather, high winds, and hazardous sea conditions (for small boats) will persist for many hours. Stormy conditions are usually the case.

Description

Stratus (St) clouds (figure 3–9) are low, gray cloud layers with rather uniform bases and tops. These dull clouds give the sky a heavy, leaden appearance. Only a fine drizzle or snow grains fall from true stratus clouds because there is little or no vertical motion in them. Stratus clouds are sometimes formed by the gradual lifting of a fog layer.

Meaning

Stratus clouds do not signify much potential danger. If the wind speed should decrease markedly when stratus clouds are present in large quantity, the base of the cloud could lower to the earth's surface, resulting in a thick fog. Other than a light drizzle, no precipitation should be expected except from higher clouds when stratus forms in advance of a warm front. In this case, the rain supersaturates the colder air below the surface of the warm front, and stratus clouds—or perhaps fog—will form.

Figure 3–9. Stratus clouds (with fractostratus present). Courtesy of R. K. Pilsbury.

Description

Stratocumulus (Sc) clouds (figures 3–10 and 3–11) are gray or whitish irregular layers of clouds with dark patches formed like rolls. These clouds frequently look like altocumulus clouds, but they are at a much lower level. They consist of water droplets, except in extremely cold weather. These clouds do not, as a rule, produce anything but light rain or snow.

Meaning

Stratocumulus clouds, which form from degenerating cumulus clouds, are usually followed by clearing at night and fair weather. The roll-type stratocumulus is characteristic of the cold seasons over both land and water, where the air is cooled from below and mixed by winds of 15 knots or higher. These clouds will persist for long periods of time under proper air-to-land or air-to-sea temperature relations. Visibility, however, can be seriously reduced in stratocumulus drizzle or snow.

Description

Cumulus (Cu) clouds (figures 3–12 and 3–13) with only a little vertical development are puffy, cauliflower-like clouds whose shapes constantly change. These

Figure 3–10. Stratocumulus clouds. Courtesy of R. K. Pilsbury.

Figure 3–11. Stratocumulus clouds (with cumulus below). Courtesy of R. K. Pilsbury.

Figure 3–12. Fair-weather cumulus clouds. Courtesy of R. K. Pilsbury.

Figure 3–13. Cumulus clouds (with vertical development). Courtesy of R. K. Pilsbury.

clouds are a brilliant white in the sunlight, often extending from a relatively dark and horizontal base.

Meaning

Cumulus clouds, when detached and with little vertical development, are termed *fair weather cumulus* (see figure 3–12). The weather is fine, and nothing hazardous is in the offing. However, when cumulus clouds swell to considerable vertical extent (figure 3–13), heavy showers are likely, associated with gusty surface winds in the vicinity of the showers. Since these clouds normally cover about 25 percent of the sky, they can often be circumnavigated.

Description

Cumulonimbus (Cb) clouds (figures 3–14 and 3–15) are heavy, dense clouds of considerable vertical extent (often to 45,000 feet and higher) in the form of a mountain or huge tower (figure 3–14). These clouds are the familiar *thunderheads*. The upper part of these clouds is usually smooth, sometimes fibrous, with the top flattened to an anvil shape or a vast cirrus plume. These clouds consist of ice crystals in the upper portion and water droplets in the lower portion.

Meaning

Cumulonimbus clouds, or thunderheads, which sometimes reach above 65,000 feet, are to be avoided if at all possible. Very gusty surface winds in the vicinity of the thunderstorm, heavy rain, lightning, frequently hail, and, in general, a bad time can be expected in the immediate vicinity of these clouds. (See figure 3–14).

Figure 3–14. Cumulonimbus cloud (with classic anvil). Courtesy of R. K. Pilsbury.

Figure 3–15. Mammatus development on underside of cumulonimbus cloud. Courtesy of R. K. Pilsbury.

If you recognize the mammatus development on the underside of the cumulonimbus cloud, as shown in figure 3–15, get out of there as quickly as possible! A tornado or waterspout could possibly develop. Figure 3–16 illustrates a "chaotic" sky, with a little bit of almost everything, and the type of situation which small boatmen should avoid by putting into port—or not leaving port.

PRECIPITATION AND OTHER PHENOMENA

As indicated in the previous chapter, precipitation can occur only if there are clouds present. However, not all clouds can produce precipitation.

Rain is the result of the condensation of water vapor in the atmosphere, falling to earth as liquid water drops or droplets. Raindrops have diameters greater than 0.5 millimeters (mm), see Table A–10 in Appendix A. And hygroscopic (condensation) nuclei must be present.

Drizzle is the other form of liquid precipitation. The drizzle droplets have a diameter of less than 0.5 mm, are much more numerous than raindrops, and reduce visibility much more than does a light rain.

Snow is the result of the sublimation (going directly from gas to solid without passing through the liquid state) of water vapor in the atmosphere into small tubular and columnar white crystals when the air temperature is below 32°F (0° C). This form of precipitation is composed of white or translucent ice

Figure 3–16. Chaotic sky. Courtesy of R. K. Pilsbury.

Figure 3–17. The swirling mass of Hurricane Carmen in the Gulf of Mexico, to the west of Cuba, as seen by one of NOAA's NESS polar-orbiting weather satellites on 3 September 1974. The southern portion of the peninsula of Florida appears in the upper right, the island of Cuba in the center right, and the island of Jamaica in the lower right of the photo. The U. S. Gulf Coast is visible in the center and upper left. Courtesy of NOAA's National Environmental Satellite Service (NESS).

crystals, primarily in complex, branched hexagonal form, and often agglomerated into snowflakes.

Snow pellets consist of white, opaque, and either round or conical ice particles, having a snow-like structure. They are about 2–5 mm in diameter and are crisp and easily crushed. They rebound when they fall on a hard surface and often break up on impact. In most cases, snow pellets occur as showers.

Ice pellets consist of transparent or translucent pellets of ice, 5 mm or less in diameter. They may be spherical, irregular, or conical in shape. They usually bounce upon striking a hard object or the ground. They make quite a sound upon impact.

Sleet is generally transparent, globular, solid grains of ice which have formed from the freezing of raindrops, or the refreezing of largely melted snowflakes, when falling through a below-freezing layer of air near the earth's surface.

Hail is precipitation in the form of balls or irregular lumps of ice. It is produced by convective clouds, almost always a cumulonimbus cloud, or thunder-

Figure 3–18. Killer-Hurricane Fifi, left center, in the Caribbean Sea and approaching the coast of Honduras, as seen by NOAA's NESS SMS/GOES weather satellite on 17 September 1974. Courtesy of NOAA's National Environmental Satellite Service (NESS).

head. A single unit of hail is called a hailstone. Hail has a diameter of 5 mm or more. Sometimes, they are the size of golf balls and larger. Thunderstorms which are characterized by strong updrafts, a large, liquid water content, large cloud drops, and extend to great heights in the atmosphere, are favorable for the formation of hail.

Frost is the covering of minute ice crystals on a cold surface. It is formed by the sublimation of water vapor directly into ice crystals at temperatures below 32°F.

Ice storms are the result of freezing precipitation. A heavy coating of clear and relatively smooth ice (glaze) forms on exposed objects by the freezing of a film of supercooled water deposited by rain, fog, or drizzle. It can cause extremely hazardous small-boat operations and conditions.

Dew results when the water vapor in the air condenses on cool objects or surfaces—especially at night and under clear skies.

Lightning, in the simplest of terms, is any and all of the various forms of visible electrical discharge produced by thunderstorms. An important effect of worldwide lightning activity is the net transfer of negative charge from the atmosphere to the earth.

Thunder is merely the sound emitted by the rapidly expanding gases along the channel of a lightning discharge. Thunder is seldom heard more than about 15 miles from the lightning discharge. On occasion, however, it has been heard as far away as 25 miles from the discharge.

4
the barometer
and its use

"There is no such thing as bad weather,
Only different kinds of good weather."
                    —John Ruskin 1819–1900

Air is one of the physical essentials of life. The condition of the air constitutes a major part of the environment of man, and its diversities are reflected in his activities and culture. Man's production increases during periods of high pressure; most suicides occur during periods of low pressure. And yet, atmospheric pressure (the "push" per unit area), one of the most important of all meteorological measurements used in weather analysis and forecasting, is probably the least detectable by man's senses.

AIR HAS WEIGHT

The atmosphere is held to the earth by the force of gravity. Therefore, it has weight which is manifested by a downward pressure. The weight of a cubic foot of air at sea level is about 1.2 ounces, or .08 pounds. The density—weight divided by volume—of air, then, is 0.08, expressed in pounds per cubic foot. Unlike water, air is highly elastic and compressible. Since the air is fluid and for practical purposes in statical equilibrium, it follows that the pressure at any point in the atmosphere is the same in all directions (downward, upward, and laterally).

A column of air from the earth's surface to the top of the atmosphere exerts a pressure on the earth's surface equivalent to a column of mercury 29.92 inches, or 76 centimeters, or 760 millimeters high, or a column of water 34 feet high.* This pressure is 1,013.25 *millibars* (units normally used on weather maps). This is why mercury, rather than water, barometers are used to measure atmospheric pressure. A 34-foot-tall water barometer would hardly be the instrument for either the seaman or landlubber.

In consequence of the weight of the air, the atmosphere exerts a pressure upon the earth's surface amounting to about 14.7 pounds per square inch (lbs/sq. in.), or about one ton per square foot. By international agreement, this is defined as *one standard atmosphere*. So we see that air is hardly a light substance. The mass of the atmosphere above a modest-size house with an area of 1,500 square feet is over 1,400 long tons! And a small boat of only 1,600 cubic feet contains over 100 pounds of air.

NORMAL ATMOSPHERIC PRESSURE

Although the atmosphere pressure varies somewhat hour-by-hour and from day to day, the *normal* value is quite well known and can be (and is) expressed in many different units. It is usually given in *pressure* units (units of force divided by area), but it is often given in terms of its balancing column of mercury or water. Table 4–1 is a list of the more common expressions for this normal atmospheric pressure.

In middle and high latitudes, barometer readings at sea level usually range between 970 mb. (28.64 in.) and 1,040 mb. (30.71 in.), but readings as high as 1,060 mb. (31.30 in.) and as low as 925 mb. (27.32 in.) may occur occasionally, and substantially lower pressure readings have been known to exist near the centers of tropical hurricanes and typhoons. Table 4–2 consists of a comparison of the various barometer scales.

*At latitude 45° and at a temperature of 32°F (0° C) around the measuring instrument.

Table 4–1. List of Common Expressions for "Normal" Atmospheric Pressure.

1,013.25	Millibars (mb)
1.01325	Bars (1 Bar = 1,000,000 dynes/cm²)
29.92	Inches of Mercury
760.00	Millimeters of Mercury
76.00	Centimeters of Mercury
14.66	Pounds Per Square Inch
33.9	Feet of Water
1,033.3	Centimeters of Water
1,033.3	Grams Per Square Centimeter
1,013,250.0	Dynes Per Square Centimeter

(Note: One millibar is 1,000 dynes per square centimeter. A dyne is the unit of force in the centimeter-gram-second system of measurements. The equivalent weight [English units] is 0.0145 pound per square inch.)

Table 4–2. A Comparison of Barometer Scales.

Inches	Millibars	Millimeters
31.00	1050.0	787.0
30.50	1032.9	774.7
30.00	1015.9	762.0
29.92*	1013.2*	760.0*
29.50	999.0	749.3
29.00	982.0	736.6
28.50	965.1	723.9
28.00	948.2	711.2
27.50	931.3	698.5
27.00	914.3	685.8
26.50	897.4	673.1
26.00	880.5	660.4

1 Inch = 33.86 Millibars = 25.4 Millimeters

(* Normal Atmospheric Pressure)

WEIGHT OF THE EARTH'S ATMOSPHERE

Since the normal pressure of the earth's atmosphere is known, it is quite simple to calculate the weight of all the air that surrounds the earth. If we multiply the pressure (in pounds per square inch) by the number of square inches in the earth's surface, the result will be the total weight of the earth's atmosphere. This gives correctly the total *weight* of the atmosphere but does not give the total *mass* quite correctly, because the outer portions have a slightly smaller weight per unit-mass than the inner portions. However, this correction is not very important.

The area of the earth in square inches is

$$A = 4 \times \pi \times r^2$$
$$= 4 \times 3.1416 \times (4{,}000 \times 5{,}280 \times 12)^2$$
$$= 8.1 \times 10^{17} \text{ square inches in the earth's surface}$$

Then, the weight of the earth's atmosphere is

$$W = 14.7 \times 8.1 \times 10^{17}$$
$$= 11.8 \times 10^{18} \text{ pounds}$$
$$(11{,}800{,}000{,}000{,}000{,}000{,}000 \text{ pounds})$$
$$= 5.9 \times 10^{15} \text{ tons}$$
$$(5{,}900{,}000{,}000{,}000{,}000 \text{ tons})$$

VARIATIONS OF PRESSURE WITH HEIGHT

As mentioned previously, atmospheric pressure is merely the weight of the column of air above a unit-area at the point in question—the column extending to the top of the atmosphere. If a unit-area at a higher elevation is considered, the pressure will be less because there is a smaller quantity (shorter column) of air above it. Thus, the pressure of the air is also a function of the elevation of the unit-area in question. The logical upper limit of the atmosphere is reached when it has become so rarified that its expansive force and the centrifugal force due to its rotation are equaled by the force of gravity holding it to the earth. The air has no very definite limit, but grows gradually thinner until it becomes imperceptible.

The atmosphere is very compressible, so the lower layers of air are much more dense than the upper layers, and for this reason, the pressure falls very rapidly with an increase in elevation through the lower layers of the atmosphere and much more slowly in the upper layers. So great is the compressibility of the air that one-half of the entire mass of the atmosphere is below 3.5 miles (19,000 ft.), and 97 percent of the mass is below 18 miles (95,000 ft.)! Thus, the top of Mt. McKinley, which has an elevation of 3.8 miles, is above more than half of the earth's atmosphere. When you're jetting about nearly 7.5 miles up (40,000 ft.), considerably *more than half* of the earth's atmosphere is below you!

As a first approximation, we can say that the pressure decreases 1/30th of its value at any given moderate elevation with an increase of 900 feet in height. Starting with a pressure of 30.00 inches (1,015.92 mb.) at sea level, at 900 feet it will have fallen to 29.00 inches (982.05 mb.). During the next 900-foot rise (to an 1,800-foot elevation), it will have fallen 1/30th of 29.00 inches to 28.03 inches (949.20 mb.); the pressure change will continue at about this geometric ratio for each successive change of 900 feet. But the density and weight of the air depend upon its temperature and, to a lesser extent, upon the proportion of water vapor in it, and the force of gravity. Consequently, no *truly accurate* correction for elevation can be made without a consideration of these factors—especially the temperature.

BAROMETERS AND BAROGRAPHS

Pressure measurements from all over the world are essential for the plotting of barometric values on weather maps (along with many other atmospheric measurements) and the analysis and drawing of *isobars*—lines connecting points of equal barometric pressure. The word *isobar* is derived from the Greek, *isos*: equal; and *baros*: weight. By drawing lines connecting points of equal pressure, weathermen determine the pressure patterns all over the world and locate, track, and predict the movements of and changes in, storms, high-pressure areas, weather fronts, low-pressure areas, wind fields, and many other items of great interest to the mariner. Pressure is measured by two types of barometers—*mercurial* and *aneroid*.

Mercurial Barometers

The mercurial barometer is the most accurate of barometers and is used by almost all weather stations around the world. Many larger ships also have a mercurial barometer on board. Mercurial barometers can be of two types, essentially, as shown schematically in figure 4–1. In figure 4–1(a), the glass tube containing the mercury is U-shaped, with one end sealed and the other end open to the atmosphere. This is known as the *siphon-type barometer*. The mercury column is supported by the pressure of the air on the open end. When the air pressure is high on the open end of the tube, the mercury column is also high. As the pres-

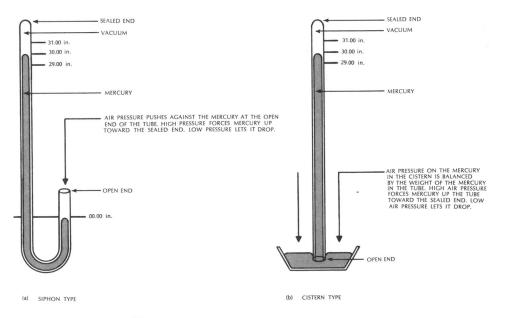

Figure 4–1. Mercurial barometers.

sure of the atmosphere decreases on the open end, the mercury column lowers. Pressure readings are made either in inches or in millibars.

The other variation of mercurial barometer, shown schematically in figure 4–1(b), is practically the same—with a few modern refinements—as the one that the Italian physicist Torricelli invented in 1643. Briefly, a long glass tube, a little over a yard long, with one end sealed and the other end open, is filled completely with mercury. The open end is then temporarily sealed (with a cork or finger), inverted, and placed in a cistern partly filled with mercury. When the cork or finger is removed, the mercury in the tube will sink somewhat and come to rest at a level of about 30.00 inches above the level of the mercury in the cistern. There is, then, a vacuum above the mercury in the tube and, therefore, no atmospheric pressure above the mercury within the tube. Since the atmospheric pressure acts on the free surface of the mercury in the cistern, it is clear that the weight of the mercury column above the free surface of the mercury in the cistern must be equal to the weight of the air column above the same surface. The length of the mercury column at any moment indicates the atmospheric pressure, and it is measured by means of a scale placed alongside the glass tube. This is called the *cistern-type barometer*. Weather station mercurial barometers are precision-made instruments and are accurate to 1/1000 inch. Figure 4–2 is a photograph of one of the latest types of mercurial barometer.

Aneroid Barometers

Basically, an aneroid barometer (a barometer without fluid), shown schematically in figure 4–3, is a corrugated metal container from which the air has been removed (a vacuum exists). The corrugations and a spring inside the container prevent the air pressure from collapsing the container, or chamber, completely. As the air pressure increases, the top of the container bends in. As the air pressure decreases, the top of the container bows out. Gears and levers then transmit these changes to a pointer on a dial.

Although aneroid barometers are not as accurate as mercurial barometers, they are much cheaper and much more robust since atmospheric pressure is balanced by elasticity forces rather than by mercury in a long glass tube. Consequently, they are ideal pressure instruments for use on board smaller sea-going vessels and boats. Because of the relation between pressure and altitude (described previously), aneroid barometers are also used as altimeters. In this case, the dial is marked in feet to show *altitude* instead of *pressure*. Figure 4–4 is a photograph of one of the newest models of aneroid barometers.

The Barograph

A barograph is simply an aneroid barometer which has been configured so that atmospheric pressure changes are *recorded* on a paper-covered rotating drum. As shown in figure 4–5, a date/time-pressure record sheet is wrapped around a cylinder or drum. Inside the cylinder is a clock mechanism which rotates the cylinder once every seven days (or some other number of days). The pen arm is actuated by atmospheric pressure changes (up and down) as the cylinder rotates, and it writes a continuous record of pressure on the graduated sheet of paper.

Figure 4–2. Precision observation barometer. Courtesy of Science Associates.

BAROMETER CORRECTIONS

For barometer readings to be made comparable around the world and to be significant, the effects of the variation of gravity between different latitudes, of height above sea level, and of temperature of the instrument must be eliminated. This is done by making small corrections to barometer readings, and the process is known as the *reduction of barometric pressure to standard conditions*.

Temperature Corrections

Mercury expands as it increases in temperature, so that the same atmospheric pressure is balanced by a longer column of mercury on a warm day and by a shorter column on a cold day. A correction is applied to obtain the reading of the barometer that it would show at the standard temperature. Depending on the particular type of barometer used, this standard is either 12°C (54°F) or 0°C (32°F).

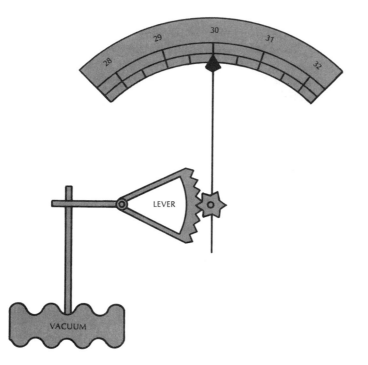

Figure 4–3. Aneroid barometer. Increasing air pressure pushes down the top of the vacuum chamber. This pulls down the attached lever and works the gears, which move the pointer to the right. Decreasing air pressure causes the pointer to move to the left.

Figure 4–4. Navy-type barometer. Courtesy of Science Associates.

Figure 4–5. Utility barograph. Courtesy of Science Associates.

Altitude Corrections

Atmospheric pressure, which is a measure of the weight of air above any point, will naturally be less on the bridge of a ship than at sea level. In order to make pressure readings all comparable around the world, the pressure that is read on the instrument needs to be corrected to obtain the pressure that the instrument would record assuming it could be located at sea level in the same place. The actual difference depends not only upon the height of the instrument, but also upon the density of the air between it and sea level. The density of the air in the column between the instrument and sea level varies according to the temperature of the air in that column, and this is obtained by reading the temperature of the air (dry-bulb).

Latitude Corrections

Because of the flattening of the earth at the poles and the fact that the vertical component of the centrifugal force due to the earth's rotation (which acts in the opposite direction from gravity) is greatest at the equator, the force of gravity increases steadily in going away from the equator toward the poles. The standard value of gravity is taken as 980.665 centimeters/second², which is practically the same as that at sea level in a latitude of 45 degrees. Thus, a *negative* correction must be applied in low latitudes, where the barometer always reads "high" by a small amount because the "weight" of the mercury is less than at latitude 45 degrees. A small *positive* correction must be applied in latitudes higher than 45 degrees.

Index Corrections

Even after all the above corrections have been made, it is found that the readings from individual barometers still differ slightly because of the differing values of capillarity of the mercury. *Capillarity* is the action by which the surface of a liquid, where it is in contact with a solid, is elevated or depressed, depending upon the relative attraction of the molecules of the liquid for each other and for those of the solid. These index corrections are determined individually for every barometer, the amount of which is clearly stated on the barometer's certificate.

Barometer Correction Example

Assume that you are aboard ship, on the bridge, at latitude 27 degrees. The height of the bridge is 53 feet, the thermometer attached to the barometer reads 77°F, the true temperature of the air is 78°F, the index correction of the barometer is +0.3 millibars, and your mercurial barometer reads 1017.3 millibars (30.041 inches). Standard conditions are 32° F(0°C); gravity of 980.665 cm/sec.². What is the true sea-level pressure reading? Refer to either the *Smithsonian Meterological Tables*, the *Marine Observer's Handbook*, or another suitable publication.

	Millibars
Uncorrected Barometer Reading	1017.3
Index Correction	+ 0.3
	1017.6
Temperature Correction for 77°F	− 4.3
	1013.3
Height Correction for 53 ft.; air temp. 78°F	+ 1.8
	1015.1
Gravity Correction in Lat. 27 degrees	− 1.6
CORRECTED BAROMETER READING	1013.5

Aneroid barometers need only to be corrected for index error and for altitude. These instruments work on the principle of the balancing of atmospheric pressure by a force due to the elastic deformation of a strong spring attached to a metal box from which the air has been exhausted and which is thereby prevented from collapsing under the external pressure of the atmosphere. Therefore, aneroid barometers do not need to be corrected for changes in gravity. They normally include a device for compensating for small changes of temperature which is ensured either by using a bi-metallic link or leaving a calculated small amount of air in the vacuum chamber. Aneroid barometers that are compensated for temperature are usually so marked.

The readings of aneroid barometers after correction, as indicated immediately above, should be compared as frequently as possible with corrected readings of mercurial barometers. The reason for this is that the index error of all aneroid barometers is likely to change quite frequently due to changes in the elasticity of the metal of the vacuum chamber. Any National Weather Service office will be glad to make barometer comparisons for you at no cost. Foreign weather offices will also render this service.

DIURNAL VARIATION OF PRESSURE

The frequently occurring changes of pressure as shown by a barograph trace may be due to many causes, and in middle latitudes in winter, it is not possible to discern any systematic variation. But in lower latitudes, and in the summer months in middle and higher latitudes, a diurnal variation of pressure with a definite pattern becomes evident. Over a long period of time, the mean daily pressure range is about one millibar in middle and higher latitudes. But in lower latitudes and in the tropics, sometimes two to three millibars are observed. Figure 4–6 illustrates a typical diurnal pressure curve in the lower latitudes, with maximum values at 10:00 a.m. and 10:00 p.m. (local time) and minimum values at 4:00 a.m. and 4:00 p.m. (local time). These result from the effects of atmospheric tides.

In middle and higher latitudes, the irregular pressure changes are usually so much larger than the diurnal variations that the latter need not be taken into account, as a rule. In lower latitudes and in the tropics, these irregular changes are usually much smaller than the diurnal change from which they can hardly be distinguished, until the known diurnal change has been subtracted. This procedure is most important in tropical regions, because a fall of pressure of two or

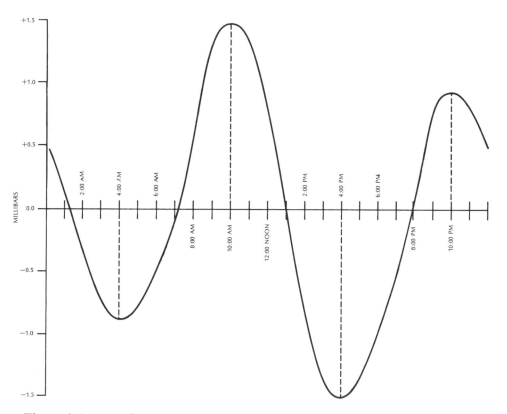

Figure 4–6. Diurnal variation of pressure in lower latitudes.

three millibars below the known value of the diurnal change of pressure may be the first indication of the generation or the approach of a tropical hurricane or typhoon. One should also be on the lookout for cirrus and cirrostratus clouds and an above-normal, long-period swell.

PRESSURE TENDENCY

The name *tendency* is given to describe the *rate of change* of pressure with time. In standard meteorological practice, it usually refers to the time interval of three hours before the pressure observation is made. It is most important for weather-men and mariners to know the change and rate of change of pressure. A ship's report of barometric tendency, of course, does not tell this pressure change directly, because part of the tendency is due to the ship's progressive movement. If the ship is at anchor or in port, the tendency is representative. However, by adjusting for the ship's direction and speed of movement, the true barometric tendency can be easily calculated, if one has the latest weather map available.

Corrected readings of barometric tendency help weathermen and mariners to determine the movements and intensity changes of the various pressure systems, since the barometer usually falls in advance of a low-pressure system (and to the rear of a high-pressure system) and rises in advance of a high-pressure system (and to the rear of a low-pressure system). If the pressure system is stationary, the tendency indicates whether the pressure system is increasing or decreasing in intensity. This will be discussed in detail in a later chapter.

Lines drawn on a weather map through points of equal barometric tendency are called *isallobars*. These isallobars are of great value to weather forecasters because they show the rate of change of pressure with respect to time all over the globe.

5

significance of high-pressure and low-pressure areas

"Look not to leeward for fine weather."
                                —J. Heywood 1546

As discussed in chapter 2, the unequal heating of the earth between the equator and the poles causes meridianal (north-south) winds. The rotation of the earth deflects the winds to the *right* in the *northern* hemisphere and to the *left* in the *southern* hemisphere. The result is that this movement of great masses of air creates the overall pattern of the earth's circulation of air parcels. But it does something else, too. It creates huge whirling masses of air called high-pressure cells, or simply *highs*, or anticyclones. It also creates low-pressure cells, or simply *lows*, depressions, or cyclones.

ISOBARS

Atmospheric pressure readings are taken simultaneously all over the world, and then these readings are *reduced to standard conditions* as described in chapter 4. When these values are plotted on charts, lines are drawn connecting points

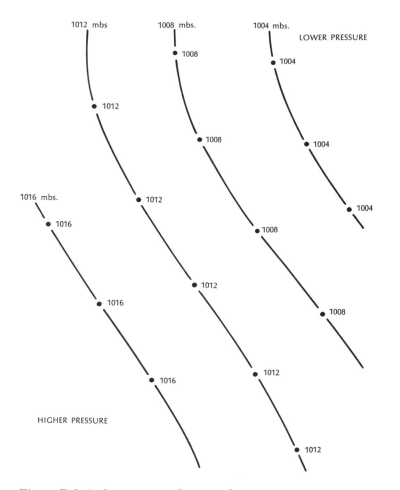

Figure 5–1. Isobars on a surface weather map.

of equal pressure. (See figure 5–1.) These lines of equal (or constant) pressure, called *isobars*, resemble equal-altitude contours which define the hills and valleys on a topographic chart, as shown in figure 5–2. A chart for a particular time which includes such data as wind, weather, temperature, clouds, isobars, and so forth, for a large number of stations is called a weather map, or more frequently a *synoptic chart*, because it gives a synopsis or general view of the weather conditions over a large area (or the entire globe) at a given instant of time. Isobars are usually drawn at intervals of four millibars (mbs.); i.e., 1012, 1016, 1020 mbs., and so forth, on weather maps. Any synoptic chart will show a distribution of pressure in which there are regions of high pressure and low pressure resembling the mountains, hills, ridges, and valleys found on topographic charts. A typical region of high or low pressure on a weather map has a rather compact shape, may be 1,000 to 2,000 miles in diameter or width, and is seldom less than a few hundred miles in width.

PRESSURE GRADIENT

The idea of *pressure gradient* is probably best illustrated by reference to a topographic contour map, where at any particular place or point we can estimate the gradient, or slope, of the ground. The steepest slope is seen to be at a right angle to the contour lines, and its numerical value is greater where the contour lines are close together. Similarly, we speak of the *pressure gradient* as being at a right angle to the isobars, its magnitude being measured by the ratio: pressure difference/distance, in suitable units. Thus, if two adjacent isobars (at a four-millibar interval) are 50 miles apart, the pressure gradient is 4/50, or 0.08 mb./mile between them, at a right angle to the isobars, and in the direction *from high toward low* pressure.

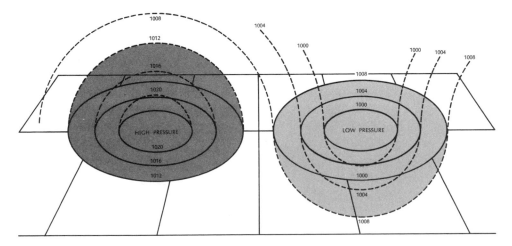

Figure 5–2. A three-dimensional view of isobars.

PRESSURE AND WIND

Fundamentally, the wind (simply air in motion) is a balance of three forces (if we exclude curvature). Referring to figure 5–3, these forces are the coriolis (horizontal deflecting) force C, the frictional force F, and the pressure force P. Arbitrarily assume that a west wind is blowing (W). The coriolis force C, we know from chapter 2, is at a right angle to the wind (and pulling to the right in the northern hemisphere). The frictional force F is opposite to the direction of the wind W. The resultant (combined effect) of the coriolis force and the frictional force is indicated by the arrow R. So, to obtain balanced motion, the pressure force P must be equal to the resultant R, but must have a direction opposite to that of R. Since the pressure force is perpendicular to the isobars (acting in the direction from higher toward lower pressure), we see from figure 5–3 that the wind W at the earth's surface must blow at an angle across the isobars, in a direction from higher toward lower pressure. This is most important to remember.

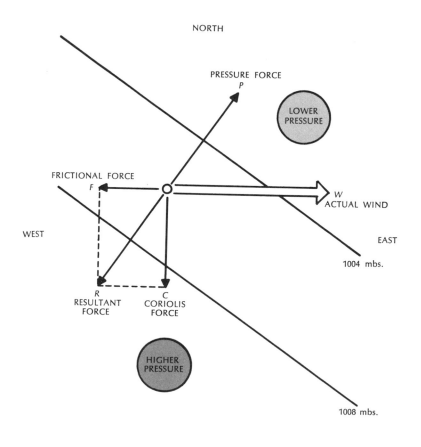

Figure 5–3. The actual wind as a balance of three separate forces.

The angle between the wind direction and the isobar increases with increasing friction. Therefore, the angle is greater over land than over water. Over land, this angle averages about 30 degrees; over water, about 15 degrees. As we ascend from the earth's surface, the frictional force diminishes and, for all practical purposes, is negligible at a height of 3,000 feet. At this level, then, the coriolis force is balanced by the pressure force, as shown in figure 5–4, and the wind blows parallel to the isobars. Thus, the surface isobars are indicative of the wind direction and speed at the 3,000-foot level. Since the frictional force decreases with increasing height, the wind velocity increases as we ascend through the atmosphere. Figure 5–5 illustrates the variation of wind speed and direction with height. Observations have shown that the surface wind speed over land is about 40 percent of the wind speed at 3,000 feet, whereas over the oceans it is about 70 percent. The surface wind over land crosses the isobars (from higher toward lower pressure) at an average angle of 30 degrees; over the water it is 15 degrees. The wind at 3,000 feet blows parallel to the isobars. (See figure 5–6.)

From what we have discussed thus far, it is obvious that if one stands with his back to the wind (in the northern hemisphere), low pressure will always be to his left, and high pressure to his right. In the southern hemisphere, the situation is reversed. This law was first formulated in 1857 by the Dutch meteorologist Buys Ballot and bears his name. This law is also known as the *Baric Wind Law.*

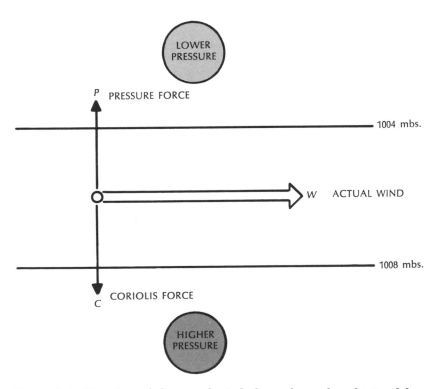

Figure 5–4. Direction of the actual wind above the surface frictional layer.

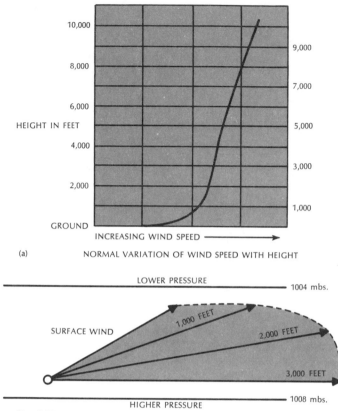

(a)   NORMAL VARIATION OF WIND SPEED WITH HEIGHT

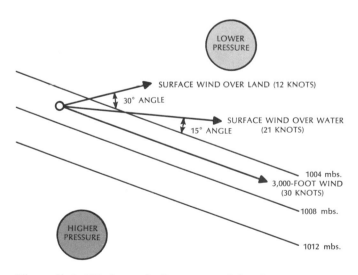

(b)   NORMAL VARIATION OF WIND DIRECTION WITHIN FRICTION LAYER

Figure 5–5. Normal variation of wind speed and direction with height.

Figure 5–6. Wind speed, direction, and height.

PRESSURE SYSTEMS

If we look at a weather map or a series of weather maps and study the isobars over a large area, we notice almost at once that there is only a limited number of types of pressure patterns (or systems) that occur in nature. Figure 5–7 shows diagrammatically the various types of pressure patterns most commonly observed on weather maps and the corresponding winds in the northern hemisphere.

The principal types of pressure patterns (or systems) are the *highs* (or high-pressure cells, or anticyclones) and the *lows* (or low-pressure cells, cyclones, or depressions). As one would expect, a *low* is defined as an area within which the pressure is low, relative to the surroundings. From figure 5–7, we see that the

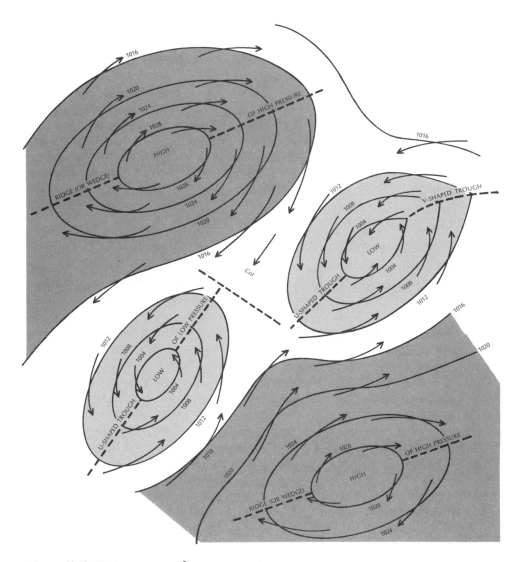

Figure 5–7. Various types of pressure systems.

wind circulation around a low is counterclockwise (in the northern hemisphere), with the wind crossing the isobars at an angle, blowing toward lower pressure.

A *high* is defined simply as an area within which the pressure is high, relative to the surroundings. The wind circulation is clockwise around an anticyclone (in the northern hemisphere), with the wind crossing the isobars at an angle and blowing from higher toward lower pressure.

A *trough* of low pressure is an elongated area of relatively low pressure which extends from the center of a cyclone. The trough may have U-shaped or V-shaped isobars (see figure 5–7). The V-shaped isobars are associated with weather fronts, which will be discussed in chapter 7. The wind circulation around a trough is essentially cyclonic (counterclockwise).

A *ridge* (or wedge) of high pressure is an elongated area of high pressure that extends from the center of an anticyclone (high). The wind circulation is essentially anticyclonic (clockwise).

A *col* is the saddle-backed region between two anticyclones and two cyclones, arranged as shown in figure 5–7.

In all the surface pressure systems, the winds blow at varying angles across the isobars, as described in the previous section, in the direction from higher toward lower pressure. Thus, the winds blow clockwise and spiral outward from the centers of anticyclones (in the northern hemisphere), while those of a cyclone blow counterclockwise and spiral in toward the center. In the very center of pressure systems, the pressure gradient vanishes, and there are either calm conditions or light, variable winds. The same is true of a col.

ANTICYCLONES

There are three basic theories regarding the formation of anticyclones (or highs), but for our purpose, arguments and counterarguments regarding the validity of these theories are not very important. Suffice it to say that local high-pressure areas may develop where air is cooled, compressed, and caused to sink from higher altitudes to the earth's surface (called *subsidence*). Figure 5–8(a), (b) shows some of the more important features of anticyclones in plan view and vertical cross-section respectively. It must be realized that since surface barometric pressure varies by only a few percent (rarely more than 3 percent in 24 hours; pressure changes in the vertical are *much* larger), air converging in one layer of the atmosphere must be nearly *compensated* by air diverging in another layer. Otherwise, surface pressure would rise or fall almost without limit.

Because the *mass* of the air is conserved, the air must move vertically from the layers of convergence to the layers of divergence. Figure 5–8(b), (d) illustrates schematically this convergent and divergent motion, and the vertical currents in highs and lows. When air converges aloft as in figure 5–8(b), it moves downward (subsides), bringing good weather with it, because the air is compressed and heated adiabatically and the relative humidity decreases markedly. When it diverges aloft as in figure 5–8(d), the upward movement of the air leads to the formation of clouds and precipitation. Since the airflow becomes increasingly counterclockwise in a convergent layer and clockwise in a divergent layer, cyclonic and anticyclonic flow must alternate with height.

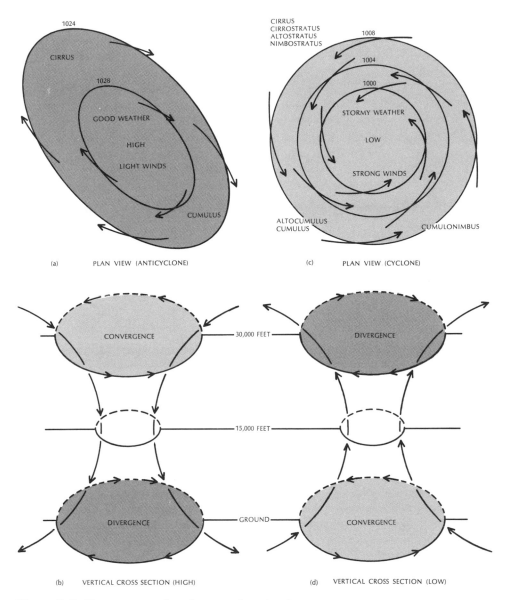

Figure 5-8. Components of cyclones and anticyclones.

Anticyclones show as much variation as people do. They range in size from the tiny ones only 200 miles across to the huge ones over 2,000 miles in length. And a "granddaddy" every now and then will cover almost the entire United States. Some highs are nearly circular in shape; others are elliptical. But near the center, almost all highs are characterized by clear skies, light wind, and good weather.

Although the speed of movement of anticyclones varies widely (and sometimes they remain stationary for days), during the summer months a good average movement would be 390 nautical miles per day (16 knots). During the winter months, the average movement of highs is somewhat greater, about 565 nautical miles per day (23.5 knots). Surface wind speeds increase considerably, going outward from the center toward the periphery of an anticyclone (i.e., the pressure gradient increases).

Anticyclones are usually classified as either *cold* or *warm* by weathermen.

Cold Anticyclones

A cold anticyclone is one in which the air at the surface and in the lower layers of the troposphere is colder than the air surrounding the high. The air in the anticyclone is thus more dense than the surrounding air, level for level. The high pressure of a cold anticyclone is therefore due primarily to the density of the lower layers of the troposphere being greater than the density of the same layers in the area surrounding the anticyclone.

Examples are the "semipermanent" anticyclones over the continents in winter. These anticyclones occur most often over Siberia and North America. They are not strictly permanent because these areas are occasionally invaded by traveling low-pressure areas (depressions). But after each period of cyclonic activity, high pressure tends to be reestablished, and may then exist for weeks with little change.

These seasonal anticyclones control the atmospheric circulation over wide areas. They are the source regions of continental polar air masses (to be covered in the next chapter). Countries that normally enjoy mild maritime conditions occasionally experience the rigors of continental winter during an invasion of continental polar or arctic air. In the British Isles, this sometimes occurs in winter months when a separate anticyclone develops over northern Europe, resulting in persistent easterly winds across the North Sea and the British Isles.

On the east coasts of North America and Asia, the outflow of continental polar air maintains a sharp temperature contrast off the coast, where a warm ocean current runs. These temperature contrasts result in the formation and development of low-pressure cells, which then travel eastward.

Although cold anticyclones are of limited vertical extent—seldom exceeding 10,000 feet—they play a very important role in the lower-level atmospheric circulation in winter. In addition to these seasonal cold anticyclones, other more transitory ones also exist.

Warm Anticyclones

In a warm anticyclone, the air throughout the greater part of the troposphere is warmer, level for level, than its environment. Near the surface, the air in a warm anticyclone therefore differs little in density from that of the surrounding air. Since surface pressure is dependent on the total mass of the air above the surface where it is measured, the excess surface pressure in a warm anticyclone must result from air of greater density than the surrounding air being present at higher levels in the atmosphere. This is brought about by the convergence of air at the higher levels, accompanied by subsidence in the lower levels. Also, in warm anticyclones, the pressure at higher levels is greater than in the surrounding air, level for level, since pressure falls more slowly with height through a column of warm air than through a column of cold air.

Examples of warm anticyclones are the oceanic subtropical belts of high pressure (the *horse latitudes* at 30 degree latitude). They are an essential feature of the general circulation and cover the areas where the air is subsiding. Consequently, the weather there is generally fine, with few clouds and good visibility. The subtropical anticyclones are the source regions of tropical maritime air masses, providing the warm air which feeds the traveling lows of the middle latitudes. The warm anticyclones tend to migrate slightly northward in summer and slightly southward in winter.

Upper-air observations made above anticyclones show that the air is relatively dry, which is to be expected in view of the subsidence taking place. They may also show an inversion of temperature (increase of temperature with height) known as a *subsidence inversion*. As a result, the highest temperature in an anticyclone is often found at a height of 1,000 to 2,000 feet above the ground. This is especially true over the sea. To some extent, this inversion aids fog formation. At sea, this happens because the moisture evaporated from the sea is kept in the lowest layers, particularly if the air stream is moving toward higher latitudes; that is, toward progressively colder sea temperatures. Over land areas, other factors, such as radiation to clear skies and high moisture content, are assisted by the absence of wind associated with the inversion to produce a drop in temperature sufficient to cause fog by condensation.

Clouds forming in the lower layers of an anticyclone, as for example, when moist air from the sea flows over land that is being heated by the sun, tend to spread out at the temperature inversion, forming a layer of stratocumulus clouds. This is more typical of the periphery of an anticyclone than of the center, where the subsidence usually prevents any cloud formation.

CYCLONES

Extra-tropical cyclones (lows of nontropical origin) frequently form along weather fronts. Hence, they occur with the greatest frequency in the higher mid-latitudes where the cold air masses and warm air masses meet along the polar and the arctic weather fronts. In the northern hemishpere, there is a maximum frequency of lows near 50 degrees north in winter, and near 60 degrees north in summer. In the Atlantic, one of the most favored regions for the development of lows is off the Virginia coast and in the general area to the east of the southern Appalachians. These lows sometimes undergo almost explosive intensification and are often called "Hatteras Storms." They move northeasterly along the Gulf Stream. Sometimes they eventually stagnate near Iceland or in the waters between Greenland and Labrador. Other times they produce dangerous storm conditions around the British Isles after having crossed the North Atlantic.

In the Pacific, there is a broad band of frequent cyclone activity extending all the way from Southeast Asia to the Gulf of Alaska. During the winter months, these storms become very intense, and usually travel northeastward to accumulate in the Gulf of Alaska. Some storms, which form on the mid-Pacific polar front, take a more southerly track and eventually reach the coast as far south as Southern California.

Figure 5–8 (c), (d) shows some of the more important features of cyclones in plan view and vertical cross-section, respectively.

Like anticyclones, cyclones exhibit a large variation in both size and shape. The smaller lows may be only a few hundred miles across, whereas the larger ones may extend to 2,000 miles or more. There have been occasions when the entire North Atlantic weather map was dominated by a single, violent low-pressure cell reaching all the way from the British Isles to Newfoundland! Some lows are very nearly circular in shape, while others are strongly elliptical.

Like highs, lows travel at varying speeds, and they, too, sometimes remain stationary for a day or two. During the summer months, lows move with an average speed of about 18 knots (432 nautical miles per day). During the winter months, they travel somewhat faster at an average speed of about 25 knots (600 nautical miles per day). Lows are much more stormy and much more sharply defined in winter than in summer.

Low-pressure systems are characterized by many types of clouds, ranging from high clouds to low clouds, moderate to heavy precipitation in the form of rain or snow (depending upon the temperature), strong winds that sometimes shift abruptly, high seas, and generally stormy conditions.

North of latitude 30°N and south of latitude 30°S, cyclones (lows) and anti-cyclones (highs) generally move from west to east. Consequently, a high- or low-pressure system located to the west of your position will *probably* envelop you at some near-future time, depending on the speed and exact direction of movement of the system and your distance from the system. Sometimes, however, these systems do not "adhere to the rule." They *may* "dissolve" before reaching your position. They *may* remain stationary for a long period of time. They *may* travel in primarily a north or south direction (instead of east)—or even retrograde toward the west. One must use the "easterly movement rule" with caution. But after reading chapter 6 (air masses) and chapter 7 (weather fronts) and acquiring an understanding of weather maps in a subsequent chapter, we shall be able to apply the rule with prudence.

FORMATION OF LOWS

Of the two major pressure systems (highs and lows), the low-pressure system is, by far, the more dangerous to mariners. Although the winds in the outer periphery of highs may sometimes reach gale force creating sea conditions undesirable or perhaps dangerous to small boating, the winds throughout an entire low-pressure system (except in the very center) and their generated waves and swell can be extremly hazardous to small boating. Consquently, all mariners should know a little something about the generation, maintenance, and characteristics of lows.

With regard to the generation and maintenance of lows, one important fact must be kept in mind: the air is spiraling inward toward the center, and some of it is being carried upward. At the top, the air is spiraling outward (normally at a faster rate than it is spiraling inward at the bottom—near the earth's surface). If this were not the case, the low would "fill," and would no longer be a low. See figure 5–9. Any theory concerning the formation of cyclones must account for these actions in the atmosphere and for the supply of energy which maintains lows and results in their movement.

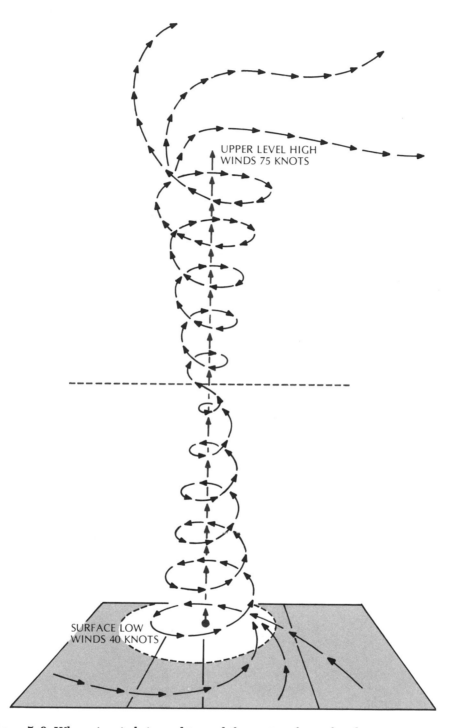

Figure 5–9. When air spirals inward toward the center of a surface low-pressure system, some of the air parcels are carried upward. At the top, (at some higher level in the atmosphere), the air spirals outward at a faster rate than it is spiraling inward at the bottom—near the earth's surface. If this were not the case, the low would "fill."

The various theories (convection, eddy, frontal, etc.) attempting to explain the formation of low-pressure systems are extremely complex and really not complete. High-speed computers will eventually provide the complete answers. But for our purposes, we're not that interested in theory. An elementary understanding of where and when lows form will suffice. An idea of the "danger signs" will serve our purpose.

Lows which develop over deserts (such as "dust devils") and other strongly convective areas are not of great interest to boatmen unless one is cruising in the Red Sea, the Persian Gulf, or similar water areas. Of greatest interest to mariners are the lows which form along weather fronts; those which form on the leeward sides of mountain ranges located near the east coasts of continents because these lows will eventually move eastward over the sea, and those generated by the combination of irregular coastlines (such as the Cape Hatteras area) and warm ocean currents (such as the Gulf Stream).

The major lows affecting mariners are those associated with weather fronts, along which masses of air of differing temperatures and densities collide and the warmer (lighter) air is forced to ascend. A "wave" is generated on the weather front, and under favorable meteorological conditions, the wave intensifies and increases in size. The low-level, inward-spiraling air parcels rise rapidly and spiral outward at greater speeds at high levels in the atmosphere, creating a low-pressure area at the earth's surface. If the intensification is of sufficient magnitude, a storm is born. This process is discussed in more detail in chapter 7. Figure 5–10 shows what a developing low looks like on a surface weather map.

Whenever you experience the combination of a rapidly falling barometer, a steadily increasing wind, increasing waves or swell, and increasing middle or low clouds, check immediately with your local National Weather Service office and/or U. S. Coast Guard office. If you're underway, get on that radiotelephone!

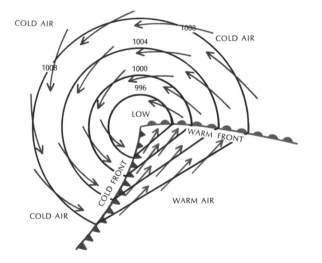

Figure 5–10. Plan view of a developing low-pressure system as drawn on a surface weather map by meteorologists.

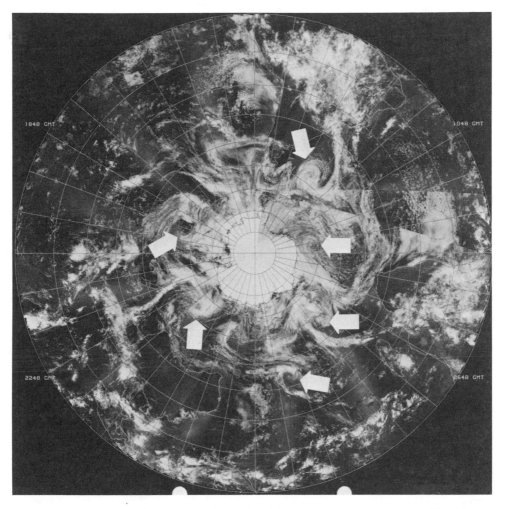

Figure 5-11. The southern hemisphere cloud patterns on 29-30 Dec. 1974. Because the rotation of the earth deflects the winds to the *left* in the *southern* hemisphere, the winds spiral inward toward low pressure in a *clockwise* manner. Note how the low-pressure cloud patterns (arrows) are reversed from northern hemisphere patterns.

HIGH AND LOW SUMMARY

	Highs	*Lows*
Weather	Generally fair. Sometimes shallow fog near center.	Usually moderate to heavy rain or snow. Stormy.
Clouds	Usually only in the periphery. Stratocumulus or cumulus in the E, and cirrus in the W.	Almost all types, ranging from high to low.

	Highs	*Lows*
Circulation	Spiraling outward in a clockwise fashion in northern hemisphere. Spiraling outward in a counterclockwise fashion in southern hemisphere.	Spiraling inward in a counterclockwise fashion in northern hemisphere. Spiraling inward in a clockwise fashion in southern hemisphere.
Winds	Light near center, increasing going outward toward periphery.	Strong everywhere, except in the very center.
Temperature	Warm or cold for relatively long periods of time, with little or no change.	Usually cold, or warm changing to cold. If of tropical origin, very warm.
Average Movement	Winter: 23.5 knots. Summer: 16.0 knots.	Winter: 25.0 knots. Summer: 18.0 knots.
Size	From 200 to 2,000 miles.	From 200 to 2,000 miles.
Shape	Circular or elliptical, and in between.	Circular or elliptical, and in between.
Marine Threat	None, except for possibility of fog near center or in periphery where warm air travels over cold water.	Yes, strong winds and high seas.

6
air masses

"No weather is ill
If the wind be still."

— W. Camden 1623

Meteorologists define an *air mass* as "a widespread body of air, the properties of which can be identified as (a) having been established while the air was situated over a particular region of the earth's surface (air mass source region), and (b) undergoing specific modifications while in transit away from the source region."[*] An air mass is also defined as "a widespread body of air that is approximately homogeneous in its horizontal extent, particularly with reference to temperature and moisture distribution; in addition, the vertical temperature and moisture variations are approximately the same over its horizontal extent."[*] Put another way, we can say that an air mass is a huge dome, or "glob," of air (enveloping many thousands of cubic miles), in which the temperature and humidity (or moisture) conditions in a horizontal plane are very similar. See figure 6–1.

The various areas within an air mass must be of common origin and their "life history" must be basically the same. A large air mass, for example, remaining over a warm ocean area for a long period of time will acquire properties of temperature and moisture which are characteristic of that warm ocean area. When that warm air mass moves toward a colder land region, it will modify its properties and develop weather phenomena which are determined by the warm oceanic origin of the air mass and its movement from a warm to a cold region. Or, a large mass of air remaining over the snow-covered arctic regions for a long period of time will acquire physical properties which are characteristic of that region.

[*]American Meteorological Society, *Glossary of Meteorology*, 1959, p. 18.

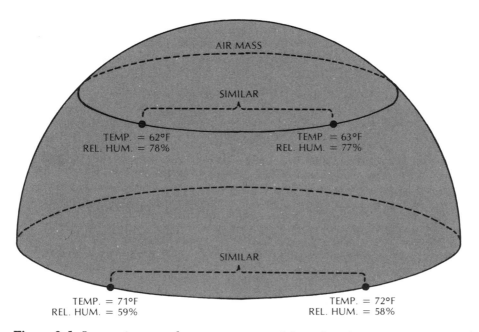

Figure 6–1. In an *air mass*, the temperature and humidity (moisture) conditions in a horizontal plane are very similar.

When that cold, dry air mass moves toward warmer regions, it changes its properties and develops weather phenomena which are determined by the arctic
source of the air mass and its movement from a very cold region toward a warm
region.

When air masses are classified according to their origin and development,
they appear as different and distinct "types," and this is of great help in attempting to forecast the weather.

As we discussed in a previous chapter, the atmosphere is almost transparent
for the sun's high-temperature radiation. It absorbs only a very small portion of
the direct solar radiation. The earth's surface, on the other hand, is a good absorbing medium, and absorbs a large portion of the sun's radiation. As a result,
the earth's surface may be considered the direct source of the primary supply of
heat for the lower atmosphere. And in any study of how huge air masses are
formed, the properties of the earth's surface are of great importance.

One method by which heat is conducted from one part of the earth's atmosphere to another is by convective currents and turbulence. This is in a vertical
direction. And this exchange of heat is especially important in the lower layers
of the atmosphere and near the earth's surface. Heat is also transported in the
atmosphere (in a horizontal direction) from one place to another by a process
called "advection,"—by large-scale air currents. Since the large-scale air currents
are primarily horizontal, the transfer of heat by *advection* is mainly in the *horizontal* direction. On the other hand, *convective currents* and *turbulence* transport heat mainly in a *vertical* direction. While the exchange of heat in the atmosphere through turbulence tends to "smooth out" temperature differences, the
horizontal transport (advection) of air may create or destroy temperature differences. And this depends upon whether the large-scale air currents flow toward
each other (convergence) or flow away from each other (divergence). Large-
scale *diverging* air currents are very important in the generation of *air masses*.
And large-scale *converging* air currents are important in the formation of *weather
fronts*. (See figure 6–2).

DESIGNATION AND TRANSFORMATION OF AIR MASSES

The general circulation of the earth's atmosphere tends to produce huge air
masses with conservative properties that are more or less uniform within each
air-mass source region. Consequently, if one travels from one air mass into another, one would experience rather abrupt changes in air-mass characteristics (or
properties) and weather conditions. If the physical properties of an air mass were
truly "conservative," air masses could be easily identified by these properties
after they leave their source region. What happens, however, is that when an air
mass leaves its source region and begins to travel, it begins to modify or change
its properties somewhat. For example, an air mass from one source region may
travel across the surface of the earth and envelop another type of source region.
When this happens, the air mass modifies its original characteristics and adopts
some of the properties of the second source region. If the air mass gets underway
again, the modified properties are further modified by the surface (land, water,

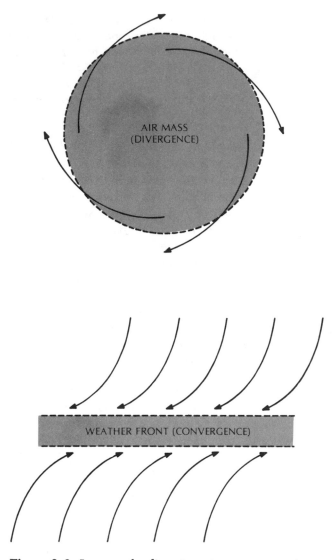

Figure 6–2. Large-scale *diverging* air currents are important in the production of *air masses*. Large-scale *converging* air currents are important in the formation of *weather fronts*.

snow, or ice) over which the air mass travels. A meteorologist is interested primarily in the last source region with which the air mass was associated and its subsequent path of travel before reaching his forecast area. He is interested in the properties which the air mass acquired in the last source region and the changes in properties which developed en route to his area. This is because the weather phenomena which develop (and which he must forecast) depend

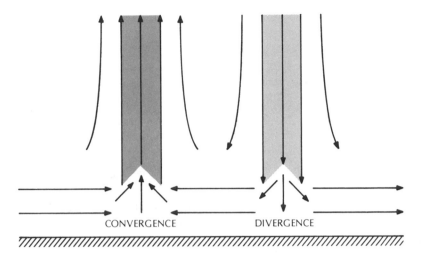

Figure 6–3. *Ascending* air currents result from *convergence*. *Descending* air currents result from *divergence*.

almost entirely on the *life history* of the air mass approaching his area. And in studying the life history of an air mass, one must consider the *source region* from which the air mass obtained its original characteristics, the *route* it traveled, the *nature* of the *underlying surface*, and the *time* it spent en route.

The concept of air masses was first introduced operationally into the science of meteorology by Professor T. Bergeron almost 50 years ago. It was he who first set forth the fundamental principles of air-mass theory, resulting in a dual classification of air masses: (1) *geographical*, comprising four principal-source air masses, and (2) *thermodynamic*, depending upon whether the air mass is warmer or colder than the underlying surface. Table 6–1 summarizes Professor Bergeron's geographical and thermodynamical classification of air masses.

Each of the four primary types of air masses (Arctic, Polar, Tropical, Equatorial) is further subdivided as continental (c) or maritime (m), depending upon whether the source region of the air mass was a continental or oceanic area. Thus, a polar air mass which formed over a continent is labeled "cP" on a weather map. A polar air mass that formed over the oceans is labeled "mP." And a tropical air mass that formed over a maritime region is labeled "mT." A third letter is suffixed to indicate whether the air mass is colder (k) or warmer (w) than the surface over which it is moving. This is very important for meteorologists to know in order to forecast the weather. If a continental polar air mass moved over a surface warmer than itself, the air mass would be labeled "cPk." If a maritime tropical air mass traveled over ground colder than itself, it would be labeled "mTw" on a weather map. There are several different air-mass classification systems and many variations used by weathermen around the world. For our purposes, however, the system described above will be of sufficient detail.

Table 6–1. Professor T. Bergeron's Geographical and Thermodynamical Classification of Air Masses.

GEOGRAPHICAL

Air Mass	Symbol	Remarks
Arctic	A	Typical of the anticyclones (high-pressure cells) over the arctic fields of ice and snow.
Polar (or subpolar)	P	Typical of the subpolar anticyclones.
Tropical (or subtropical)	T	Typical of the subtropical anticyclones.
Equatorial	E	Typical of the equatorial area of the trade-wind "belts."

THERMODYNAMICAL

Cold	k	Colder than the underlying surface. Absorbs heat from below. "Unstable."
Warm	w	Warmer than the underlying surface. Gives off heat to the underlying surface. "Stable."

HEATING AND COOLING OF AIR MASSES

The amount of solar radiation absorbed by the earth depends primarily on its color. This is called the earth's *albedo.** Of the amount of radiation absorbed, part is used to evaporate water and part is expended in chemical processes. But the major part goes to heat the earth. The heat that is absorbed is concentrated in the upper few inches of the earth, because of the very slow conduction of heat in the earth. In this way, the air that comes in contact with the earth becomes heated. As the sun climbs higher in the sky, the heating increases and the heat obtained from the earth's surface is transported to higher levels in the atmosphere. As the temperature of the earth's surface increases, the outgoing radiation also increases. About two hours after the sun reaches its maximum altitude, a balance is reached between the gain and loss of heat. The temperature of the earth has reached its maximum, and the earth then begins to cool. As one would expect, the cooling of the earth's surface has a direct effect upon the temperature of the air. As before, the influence is greatest near the ground. The result is that the air

*The reflective power of the earth is called its *albedo*. Snow and clouds reflect as much as 80 percent of the sun's radiation. Water and black soil reflect only a small part of the incoming radiation.

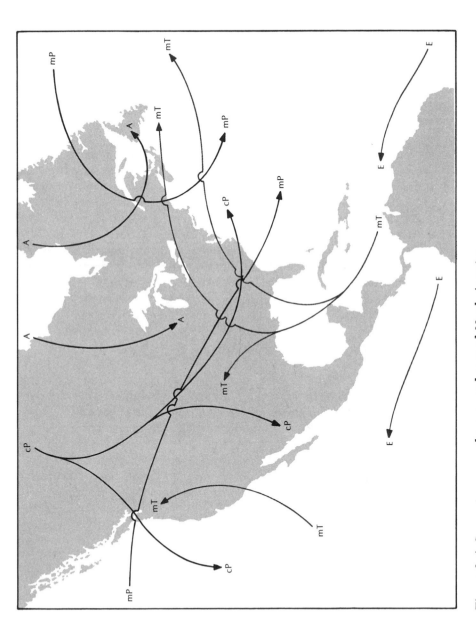

Figure 6-4. Some air mass paths over and around North America.
A = Arctic   cP = continental Polar   mP = maritime Polar   mT = maritime Tropical   E = Equatorial

cools much more quickly near the earth's surface than at high levels in the atmosphere. The air becomes "stable" again. It is no secret that the temperature over macadam roads and sandy deserts is much higher than over a field of grass, and that it is much lower over a forest area and lower still over water. Also, that temperature differences give rise to convective currents.

The situation is different over the oceans, lakes, and rivers. Although the incoming radiation is absorbed to a large degree by bodies of water, the temperature of the water surface remains almost constant both day and night. This is because of several factors. A part of the heat absorbed is used in evaporating water. The remaining heat is distributed over a deep layer of water. Although most of the radiation is absorbed in a relatively shallow surface layer, the mixing caused by the wind and waves distributes the heat in a deep layer. The result is that there is little variation in the temperature of the sea surface. And the air that is in contact with the water surface adapts its temperature to the sea surface temperature. This "regulating influence" of the ocean surface on the air temperature decreases with increasing height above the ocean surface. A few hundred yards above the ocean surface, the temperature variation is "controlled" primarily by radiation. At this altitude, the air temperature changes much more between day and night than close to the ocean surface. And at higher altitudes, the effect of radiation decreases because of considerably less water vapor (the great "absorber" in the earth's atmosphere). Over the oceans, the lower part of the earth's atmosphere tends to be stable by day and unstable at night. At greater heights in the atmosphere, the reverse is the case. This "diurnal variation" of stability and instability in the lower layers of the atmosphere over the oceans is the opposite of conditions over land. Over land, the instability phenomena (such as showers and thundershowers) occur most frequently and with greater intensity in the afternoon. Over the oceans, the diurnal variation is at a minimum and the instability phenomena frequently occur at night.

An air mass that is colder (k) than the surface over which it is moving is called a cold air mass. It is heated from below, and as it continues to travel over a warmer surface, instability develops in the lower layers and spreads upward in the atmosphere. The vertical (convective) currents which result from the instability carry heat and moisture to higher levels. These air mass changes are very similar to those in an air mass that is heated by the sun's radiation over land. If this kind of an air mass travels over a land surface, it will have the diurnal temperature changes superimposed on the effect of its movement toward warmer areas. The instability of the air varies during the day. It has a maximum during the afternoon and a minimum in the early morning hours. If the temperature difference between the air and the surface is large, the air will probably remain unstable both during the day and night. If the difference is small, the air will probably become stable during the nighttime hours.

An air mass that is warmer (w) than the surface over which it is moving is called a warm air mass. By continued movement toward colder regions, it will be cooled from below and will become extremely stable in the lower layers. The stability suppresses vertical currents (convection) and the cooling is limited to the lower layers. If this air mass then moves over a land surface, it will have the diurnal variation in stability superimposed on the effect of its movement toward colder areas. If the difference in temperature between the air and its

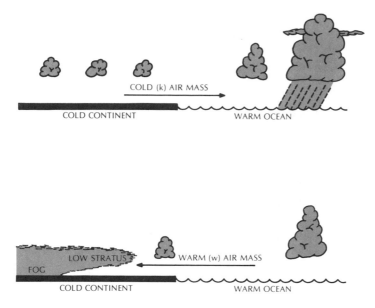

Figure 6–5. Illustrations of what happens when air masses are heated or cooled from below. Arrows show direction of movement of the air mass. In the upper diagram, the air mass is heated from below, resulting in showers. In the lower diagram, the air mass is cooled from below, resulting in low stratus clouds and fog.

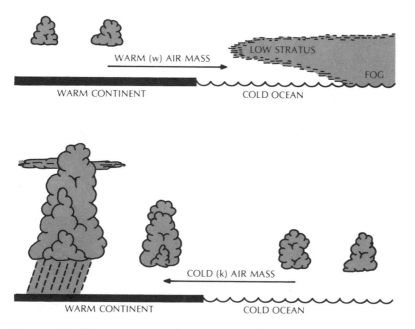

Figure 6–6. Illustrations of what happens when air masses are heated or cooled from below. Arrows show direction of movement of the air mass. In the upper diagram, the air mass is cooled from below, resulting in low stratus clouds and fog. In the lower diagram, the air mass is heated from below, resulting in showers.

Table 6–2. Weather features of Cold (k) and Warm (w) Air Masses.

Air Mass Suffix	Cloud Types	Visibility	Surface Winds	Rel. Humidity	Thermodynamically	Weather Phenomena
Cold (k)	Cumuliform	Good to Excellent	Gusty	Usually Low	Unstable	Showers, Squalls, or Thundershowers
Warm (w)	Stratiform	Fair to Poor	More or less Steady	Usually High	Stable	Fog, Light Rain, or Drizzle

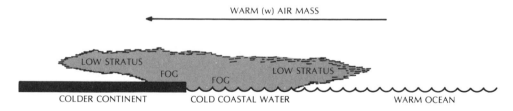

Figure 6–7. Illustration of what happens when a warm air mass travels over cold coastal water and a colder continent. The arrow shows the direction of movement of the warm (w) air mass. The air mass is cooled from below, resulting in low stratus clouds and fog. This situation can be very dangerous to small boats. It happens quite frequently off the west coasts of continents.

underlying surface is large, the air will remain stable both day and night. If the difference is small, the air will remain stable at night but will become unstable during the daytime. Air-mass conditions are different over the oceans primarily because the diurnal variation of stability is small, as compared to that over a land surface. Consequently, over the oceans, stability and instability are decided mainly by the movement of air masses, and there is only a very small diurnal variation. This is especially true in middle and high latitudes.

Two important things to remember are that a warm (w) air mass is cooled from below and becomes stable. An example of stability is low stratus-type clouds, drizzle, and fog. A cold (k) air mass is heated from below and becomes unstable. Examples of instability are gusty winds, cumulus-type clouds, showers, squalls, and thunderstorms.

THE AIR MASSES IN WINTERTIME

Arctic (A)

Over the Arctic Ocean, the air temperature near the earth's surface never reaches the extremely low values observed in northern Canada or in Siberia. This is because of the conduction of heat through the arctic ice from the warmer water underneath. Arctic air, however, is very cold at higher levels in the atmosphere and extends to great heights. Sometimes it extends all the way up to the tropopause (the zone of transition between the troposphere and the stratosphere, at an average height of 24,000 feet [8,000 meters] above the north polar region). In genuine arctic air, the specific humidity (ratio of the mass of water vapor to the total mass of the system) is lower than in cold, continental polar air. When moving out of its source region, arctic air is sometimes modified rather rapidly.

Continental Polar (cP)

In the formation of a cP air mass, the air is cooled from below. And because of the outgoing radiation, the cooling at the surface is most intense where the winds are diverging. Here, the descending air currents (subsidence) dissolve the clouds and permit the surface to radiate easily. The divergent flow of air from the central part of the anticyclone spreads the cooled air over large areas, while the subsiding air aloft which compensates for the low-level outflow, is heated by compression. The final result is an extremely cold layer of air along the earth's surface with relatively warm air aloft. The temperature conditions that gradually develop depend upon the cooling of the surface, the turbulent transfer of heat in a vertical direction, and subsidence (descending air currents wherein the air is heated *adiabatically*—no transfer of heat or mass across the boundaries of the system; i.e., *compression* results in *warming, expansion* results in *cooling*). And the amount of radiative cooling depends very much upon the temperature of the air at higher levels.

The moisture content (humidity) of the air adapts itself to the temperature so as not to exceed saturation. The higher the temperature of the air, the more water vapor it can "hold." In the lower layers, moisture is taken out of the air by frost deposited on the ground or on cold objects, or by the formation of ice-crystal fogs. In this way, the air in the central parts of anticyclones, where the winds are light and variable, is gradually desiccated, and the dried and cooled air is spread over very large areas by the diverging air currents. Because of the subsidence and the turbulent transfer of heat, the air throughout the entire column of air becomes drier and colder.

The vertical extent of cP air masses depends greatly on topographical conditions. Almost all cP air mass source regions are encircled by mountain ranges; i.e., the North American ranges, the Scandinavian ranges, the Balkan ranges, the eastern Asiatic ranges, and the Himalayan ranges. When an air mass source region is limited by extensive mountain ranges, the cold air produced is prevented from streaming out of its source region until it has attained a considerable height. Also, air from adjacent source regions of a different nature are prevented from streaming into that source region. The most extremely cold and specifically dry air is found in northern Siberia and in northern Canada. See figure 6–8.

Under favorable conditions, such as when the ground is covered with snow, cP air forms in the continental anticyclones over central Europe. These air masses are quite shallow and seldom extend higher than 6,000–7,000 feet. In this situation, and when the upper-air currents blow toward the west, central Europe is invaded by frigid cP air from eastern Europe.

The time required for a cP air mass to form depends to a great extent on the air temperature and cloudiness at higher levels in the atmosphere. Under cloudy conditions, the cooling of the underlying surface by radiation is slowed down considerably, and the air mass approaches "normal" conditions very slowly. Also, if the advection of warm air is taking place at higher levels in the atmosphere, the radiative cooling of the surface is at a slow rate. Under these circumstances, it takes several days for a cP air mass to form. But when the air is clear or there are very few clouds present, the underlying surface cools rapidly. If, at the same time, the wind velocity is high enough to produce turbulence, the entire column of air is cooled rather quickly. In this case, a cP air mass is formed in less than two days.

Of the many processes and influences which exist in the earth's atmosphere, there are five which have the greatest effect in changing air mass source proper-

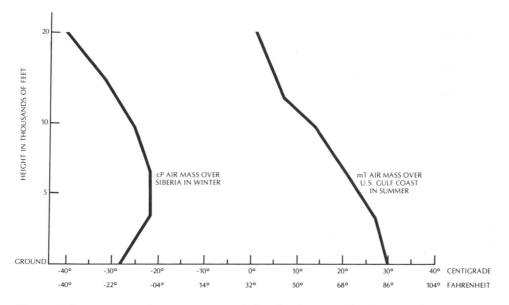

Figure 6–8. Variation of temperature with height (lapse rate) of a typical cP air mass over Siberia in winter, and of an mT air mass over the U. S. Gulf Coast in summer.

ties. These are: (1) the supply of heat from the underlying surface, (2) the supply of moisture from the underlying surface, (3) the turbulent transfer of heat and moisture in a vertical direction, (4) the effect of descending air currents (subsidence), and (5) the release of the latent heat of vaporization (about 598 calories per gram at 0°C). (As a matter of interest, the latent heat of fusion is about 80 calories per gram, and the latent heat of sublimation is about 677 calories per gram at 0°C.) The characteristics of an air mass that develops from a cP source region and moves toward a warm region depend upon which of the foregoing five influences are the predominant ones. Continental Polar air traveling toward a warmer region does *not necessarily* result in an unstable air mass.

When cP air passes a coastline and moves over warmer water, it absorbs heat and moisture in the lower levels and becomes unstable. Because of its low relative humidity, cP air frequently has few clouds inland or near the coast. But at a distance of only 100–150 miles off the coast, the cP air can have towering cumulus clouds and thunderheads. This is important to remember when heading out to sea.

Maritime Polar (mP)

This type of air mass is formed primarily in the northeastern Atlantic and northeastern Pacific oceans. Because of the small geographical extent of these regions and the normally strong wind circulation in these regions, the air rarely has enough time to acquire the typical properties to their full extent. Most of the time, the air masses occupying these regions are in a state of transformation. And depending upon the path of movement of the air mass, and the time during which the air mass has been in contact with the ocean, a great variety of conditions is observed.

As a general rule, we can say that mP air masses develop in "transitional" zones by the drawing together (concenescence) of air masses from nearby air air mass source regions. As a result, mP air exhibits a great variety of conditions. Sometimes an mP air mass resembles mT air. Sometimes it approaches the characteristics of transitional cP or transitional A (Arctic) air.

Continental Tropical (cT)

There is only one source region of cT air in winter—North Africa. In southern Asia, the NE monsoon prevents the air from becoming stagnant and absorbing the properties of the underlying surface. And because of the converging coastlines of North America, there is not enough continental area to form a source region for cT air. Except for the coastal regions which are under a maritime influence, most of Mexico and Central America are high plateau or mountain country— quite cool in winter. The cT air of North Africa is extremely dry at the ground and in the lower levels. It is moderately warm and stable. This cT air in winter is the main supply of heat for the low-pressure systems that transit the Mediterranean. The weather front that forms during the winter months in the Mediterranean is the zone of transition between the cP air from Europe and western Asia to the north, and the cT air from North Africa to the south. Rarely does the cT air invade Europe. When it does, it is at high altitudes, with little effect near the earth's surface.

Maritime Tropical (mT)

These air masses are formed almost continuously between latitudes 20°N–40°N in the Atlantic and between latitudes 12°N–40°N in the Pacific, whenever an anticyclone "settles down" in these oceans. The vertical structure of mT air masses is characterized by three basic factors: (1) the warmth and oceanic conditions of the source regions, (2) the anticyclonic subsidence in the atmosphere resulting from the diverging outflow at and near the surface, and (3) the "cellular movement" within the subtropical high-pressure cells.

As Atlantic mT air arrives in Europe, it is very stable in the lower layers. Temperature *inversions* (increase of temperature with height) occur frequently. The same is true of Pacific mT air that invades the west coast of the United States. The mT air that invades the U.S. west coast at high latitudes has had a long path over a colder underlying surface since it left its source region in a subtropical anticyclone. Consequently, its original instability has disappeared because of the cooling from below and the upward flux of moisture.

Of all the air masses, mT air is the most homogeneous of those that invade middle and high latitudes. Also, its characteristics vary very little from one case to another. Perhaps surprisingly, very little precipitation is released within mT air masses. The downpours and torrential rains usually associated with mT air occur when the mT air mass is lifted over a weather frontal surface, over an orographic barrier (mountain), or when involved with a strong cyclonic circulation (and converging air currents) such as a hurricane or typhoon.

THE AIR MASSES IN SUMMERTIME

Arctic (A)

The coldest air masses in summer are those that form over the arctic fields of ice and snow. These air masses are so shallow, however, that they lose their characteristics as they travel southward away from the source region. This is especially true from mid-June to early September. It is then very difficult to recognize the arctic characteristics of the air mass after it has traveled south of latitude 60°N.

Maritime Polar (mP)

In the summertime, there is a general outflow of air from the arctic region. This ice- and snow-cooled air gradually becomes transformed into mP air in both the North Pacific and North Atlantic oceans north of about latitude 52°N. The mP air in these regions is quite cold and humid. During its southward movement, it is heated from below and becomes unstable to a height of about 9,000–10,000 feet. An mP air mass is also formed when warm air from the continents invades the North Pacific and North Atlantic, north of about latitude 52°N. But in this case, the air is cooled from below and becomes stable. As a result, low stratus clouds and fog occur. See figure 6–9.

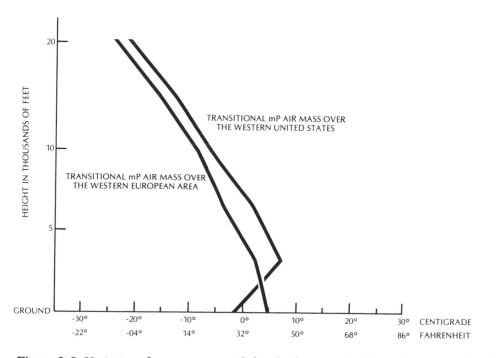

Figure 6–9. Variation of temperature with height (lapse rate) of typical transitional **mP** air masses over western Europe and the western United States in winter.

Continental Polar (cP)

These air masses form over the North American and northern Eurasian continents, north of latitude 50°N. The summer cP air masses form most frequently from the merging of A and mP air masses that invade the northern continental regions. After a long path of travel over a warm continent, the cP air gradually changes and sometimes becomes almost as warm as cT air. Summer cP air masses exhibit all variations from cold and stable A air to warm and unstable cT air. The properties of cP air vary considerably with latitude. Also, cP air has a greater annual variation of air temperature and moisture content than any other air mass. Consequently, its properties vary rapidly from month to month. In general, it can be said that cP air masses in summer: (1) have a fairly low humidity, (2) have a moderately low temperature, (3) are unstable in the lower layers, and (4) when clouds are present, their bases are usually above 2,500 feet. See figure 6–10.

Continental Tropical (cT)

Typical cT air masses in summer are produced continuously in the subtropical arid regions of the Eurasian continent, North Africa, and North America. The cT air that is formed over the deserts and steppes is extremely warm and unstable.

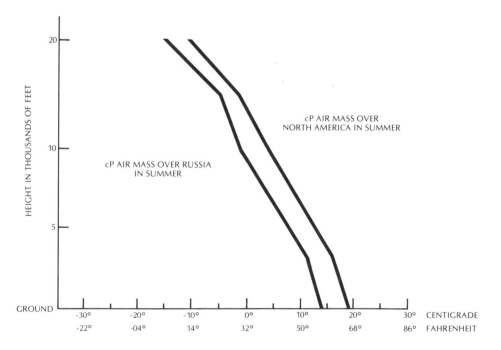

Figure 6–10. Variation of temperature with height (lapse rate) of typical cP air masses over North America and over Russia in summer.

Because of the lack of moisture (the atmospheric temperature "regulator") in the air mass, temperatures in the afternoon are very high and the nights are very cool—sometimes cold. The cT air that forms in North Africa is extremely dry. But when it travels northward, it absorbs a large amount of moisture and becomes unstable before it arrives in southern Europe. This original cT air produces the showers and thundershowers in southern and eastern Europe.

Maritime Tropical (mT)

The main source regions of mT air are the North Atlantic and North Pacific oceans between about latitudes 12°N–46°N in the Atlantic and between latitudes 15°N–45°N in the Pacific. Conditions are about the same as they were during the winter months, but the subtropical anticyclones are somewhat larger and more "permanent." Also, the ocean temperatures are higher and more uniform. In the eastern portion of the subtropical anticyclones, the northerly winds increase in summer. As a result, there is an "upwelling" of cold water along the subtropical west coasts of continents. At the same time, the ocean temperature increases to its summer maximum in the western sections of these anticyclones. The overall result is that the mT air moving toward the west on the equatorial side of the subtropical anticyclone is heated from below. And the mT air moving toward the east on the poleward side of the anticyclone is cooled from below. This is why mT air in the western part of a subtropical anticyclone is unstable, and stable in the eastern part. See figure 6–11.

Most of the rain during the summer months in the United States originates from mT air masses, either as air-mass showers or thundershowers, or as frontal rain. There is little difference between an mT air mass over the United States and one over Asia. See figure 6–8, again.

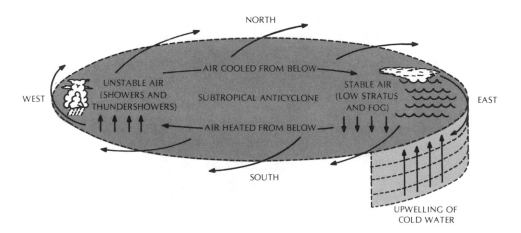

Figure 6–11. Schematic summary of some facts about, and characteristics of, maritime tropical (mT) air masses in the circulation of a subtropical anticyclone in the summertime.

Equatorial (E)

The source regions of E air are the North Atlantic Ocean, south of about 12°N, and the North Pacific Ocean, south of about 15°N. The region occupied by E air coincides with the doldrums. This air mass is extremely uniform in both time and space. It has a high moisture content, very warm temperatures, and is unstable through a deep layer. It is also the best "breeder" air mass of hurricanes and typhoons!

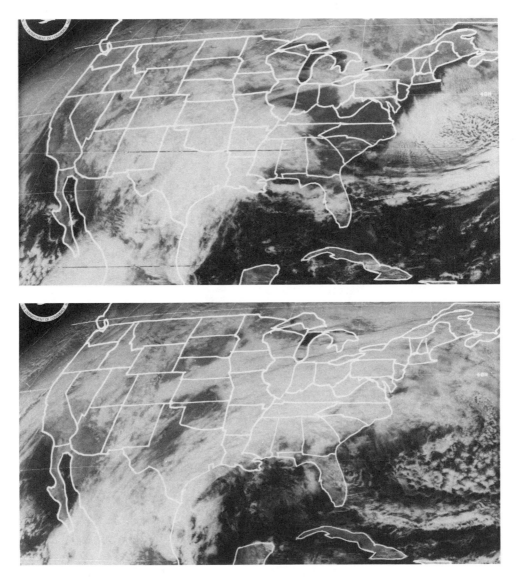

Figure 6–12. A 4-day sequence of air mass and frontal clouds and weather over the United States and adjacent water areas, as seen by the NOAA-4 polar-orbiting weather satellite. This information appears on many TV daily weather shows.

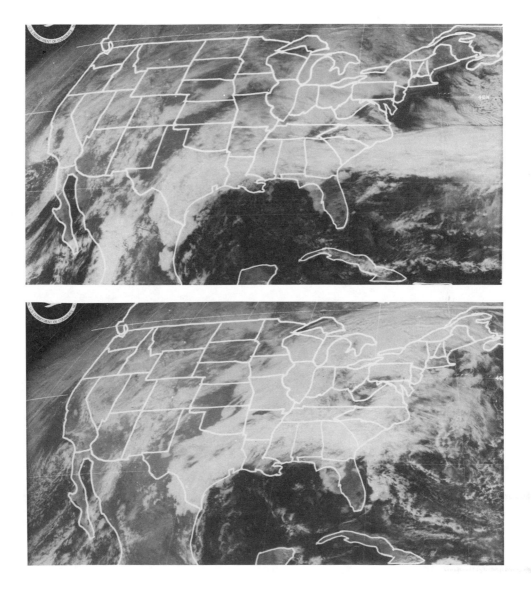

7

weather fronts and associated weather

"If it had not rained on the night of June 17, 1815,
the future of Europe would have been changed.
. . . A few drops of rain mastered Napoleon.
Because Waterloo was the finale of Austerlitz,
Providence needed only a cloud crossing the sky
out of season to cause the collapse of a world."
　　　　　　　　　　　　—Victor Hugo (1802–1885)

AIR-MASS BOUNDARIES—OR FRONTS

As we discussed in the previous chapter, although each individual air mass is quite homogeneous in a horizontal direction (at almost all levels), the boundary zones of *different* air masses can be very sharp. For example, several years ago when the writer was presenting a paper in St. Louis, Missouri, the temperature dropped abruptly from the middle 70s to the high 20s (Fahrenheit) in less than one hour with the passage of a cold front! These sharp differences across air-mass boundaries (weather fronts) exist because there is very little mixing of different air masses across weather fronts. On weather maps, the boundary zones of the different air masses are drawn in as lines, which are called *weather fronts*, or simply *fronts*. Thus, a *front* may be regarded as a line at the earth's surface, or any other nearly horizontal surface, dividing two different air masses.

The boundaries between air masses are really *zones of transition*, ranging from about 5 to 60 miles in width, but because of the small geographic scale of weather maps, fronts are usually drawn as lines on the meteorological charts.

If we exercise our mental processes and think in three dimensions, we can visualize two dissimilar air masses, or domes of air, coming close together and being separated by a surface whose intersection with any horizontal plane is the *front* as represented on that particular chart. This surface of separation between the two air masses is known as a *frontal surface*. The name *front* was first introduced during World War I by analogy with battlefronts. But the analogy goes much further, because most weather disturbances originate at fronts, and the general weather picture, as successive disturbances travel along the frontal zone, is one of constant war between two or more conflicting air masses.

The principal frontal zones around the earth in summer and winter are as shown in figure 7–1. These are as follows:

Arctic, separating arctic (A) air masses from either maritime polar (mP) or continental polar (cP) air masses.

Polar, separating continental polar (cP) air masses from maritime polar (mP) air masses, or separating either cP or mP air masses from maritime tropical (mT) air masses.

Mediterranean, separating the cold air masses over Europe during the winter months from the warmer air masses over North Africa.

Intertropical Convergence (ITCZ), referred to in older publications as either the *intertropical front*, or the *equatorial front*. This is the normally bad weather region in tropical latitudes where the northeast trade winds of the northern hemisphere and the southeast trades of the southern hemisphere either approach each other at a rather large angle, or flow almost parallel to each other. The ITCZ is not really a true front, like the fronts of middle or high latitudes, because the temperature and moisture conditions on both sides of the ITCZ are very nearly the same. This zone is one of the favorite breeding areas of hurricanes and typhoons, which will be discussed in chapter 8.

There are two definitions that we must be clear on, at the outset. A *frontal surface* is an inclined narrow layer of transition in which the meteorological (weather) elements vary abruptly. And a *front* (or weather front) is the line or zone of intersection between a frontal surface and a horizontal plane (such as the earth's surface). The layer of transition which is the frontal surface and the zone of transition which is the weather front are usually so narrow that their widths

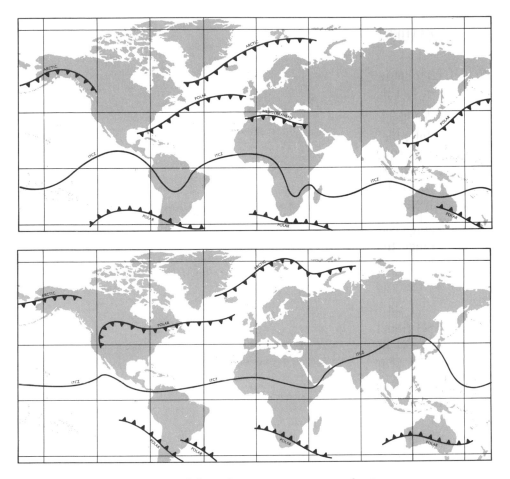

Figure 7–1. Mean positions of frontal zones in summer and winter.

are negligible when compared to the dimensions of the air masses themselves. Whether a frontal zone appears on a weather map as a perfect discontinuity or as a zone of transition depends primarily upon the scale of the map. In most cases, the scale of the weather maps is such that even fairly broad zones of transition appear as distinct discontinuities, and are drawn as a line by meteorologists, rather than as a zone. But when weathermen draw vertical cross sections, the vertical scale is so exaggerated as compared to the horizontal scale that frontal surfaces appear as zones of transition instead of a sharp discontinuity (drawn as a line).

SOME GENERAL CHARACTERISTICS OF ALL FRONTS

Frontal surfaces in the atmosphere are discontinuities in temperature (the temperature changes abruptly going from one air mass, through a frontal surface, into another different air mass). Consequently, frontal surfaces are also discontinuities in density (density is the ratio of the mass of a substance to the volume it occupies), because colder air is more dense (heavier) than warmer air. It is a well-known fact that the surface of separation between a more dense and a less dense fluid (oil and water, for example) tends to become horizontal. If the earth did not rotate, the surface of separation would be precisely horizontal when the fluids were in equilibrium. Because of the earth's rotation, however, frontal surfaces are in equilibrium when they are inclined slightly to the horizontal. The angle at which frontal surfaces are inclined to the horizontal (to the earth's surface) depends upon the wind distribution as well as the density discontinuity. Under average conditions, the slopes (inclinations) of frontal surfaces in middle latitudes range from about 1/50 to 1/200 and are as shown in figure 7–2. Although the inclination of frontal surfaces is not large, it is a very important factor in weather analysis and forecasting.

If frontal surfaces were perfect discontinuities, temperature-height curves (lapse rates) from the ground upward through the earth's atmosphere would look like figure 7–3a. But perfect discontinuities are never found in the atmosphere. Narrow layers of transition between the cold and warm air masses are actually the case because of the mixing that takes place in the atmosphere. The temperature-height curve associated with a well-developed frontal surface looks like figure 7–3b. In the case of a weak frontal surface, the associated lapse rate would have the appearance of figure 7–3c.

Atmospheric pressure on the ground or at any level in the atmosphere is equal to the weight of the column of air above it. Since cold air is more dense (heavier) than warm air, the weight of the column of cold air in figure 7–4 is greater than the weight of the column of warm air which is of the same height. Consequently, atmospheric pressure at the bottom of the column of cold air is greater than that at the bottom of the column of warm air.

Lines of constant pressure drawn on weather maps are called *isobars*. These lines are usually drawn on surface weather maps at intervals of four *millibars* (a pressure unit convenient for reporting atmospheric pressures, discussed in

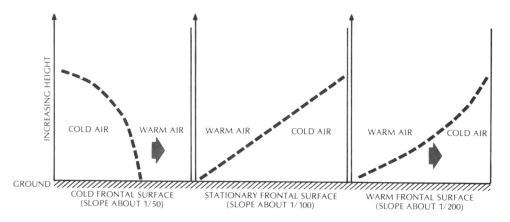

Figure 7–2. Average inclination (or slope) of frontal surfaces. The vertical scale is exaggerated compared to the horizontal scale in order to emphasize the slope.

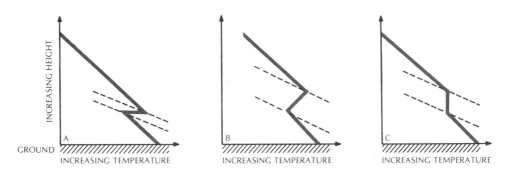

Figure 7–3. Temperature-height curves (lapse rates) associated with frontal surfaces: A, a perfect discontinuity; B, a strong or well-developed frontal surface; C, a weak frontal surface.

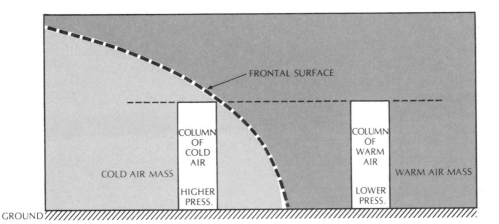

Figure 7–4. Atmospheric pressure at the bottom of the column of air in the cold air mass is greater than the pressure at the base of the column of warm air because cold air is more dense (heavier) than warm air.

chapter 4). Referring to figure 7–5, this drawing represents a small part of a surface weather map and shows a cold front moving from the NW toward the SE and the isobars associated with the front. If we trace along the 1008-millibar isobar from east to west in the warm air and toward the cold front, trace through the cold front and into the cold air, the pressure must increase because cold air is heavier (more dense) than warm air. If we trace far enough, we intersect the 1012-millibar isobar. This illustrates that *the isobars at a weather front must be refracted so that the "kink" in the isobar points away from low pressure and toward high pressure.* This is true in the case of all kinds of fronts. From figure 7–5, we see that as a weather front approaches your position, the pressure will fall. And after the front has passed, the pressure will rise. As the weather front passes your position, there will be a "kink" in the pressure trace, pointing downward. The pressure traces can have various patterns, as shown in figure 7–6, and

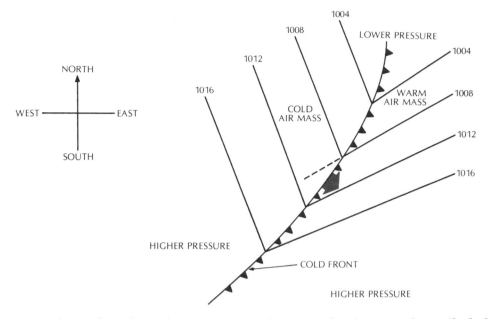

Figure 7–5. Isobars (lines of constant pressure) at a weather front are always "kinked" away from lower pressure and toward higher pressure.

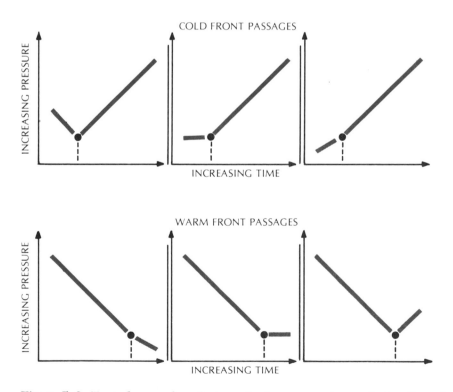

Figure 7–6. Typical examples of atmospheric pressure traces during frontal passages. The "dots" indicate the time of the frontal passage.

the kink will always point downward. As a general rule, the more intense the weather front and the greater its speed of movement, the sharper will be the kink.

As discussed in chapter 5, the winds near the earth's surface blow at a slight angle (15°–30°) across the isobars in the direction from higher toward lower pressure. As a result, the wind direction in the vicinity of a weather front must conform with the refraction of the isobars. Referring to figure 7–7, we see that a weather front is a *wind-shift line*. Weather fronts move in the direction of the wind component perpendicular to the front. So, we can formulate another rule that applies to all kinds of fronts: *if you stand facing the wind in advance of a front, the wind will shift to starboard (your right-hand side) as the front passes.* This is very important to remember! The speed of the wind depends upon the pressure gradient. The closer together the isobars, the stronger the wind. The farther apart the isobars (the weaker the pressure gradient), the weaker the wind.

The passage of a strong weather front is often accompanied by very gusty winds and squalls which are dangerous to boating operations. The gustiness and squalls depend primarily upon three factors: (1) wind speed, (2) air mass stability, and (3) vertical velocities at the weather front. Squalls can be expected during a frontal passage when the winds behind a front are stronger than the

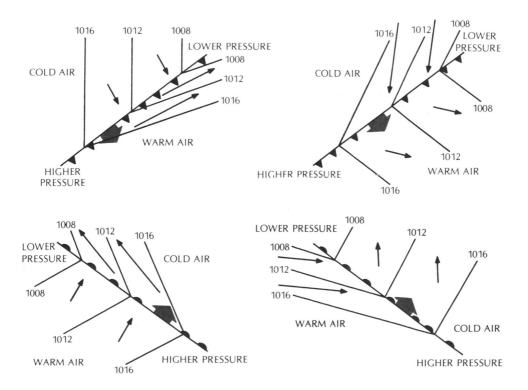

Figure 7–7. Typical isobar patterns associated with weather fronts. Note that fronts are wind-shift lines on weather maps.

winds ahead of the front. These squalls are always associated with a *veering* wind (a wind that shifts clockwise, as from SW to W to NW). (A *backing* wind is one that shifts counterclockwise, as from SW to S to SE.) Strong gusts also are likely to occur along weather fronts where rapidly moving unstable air is replacing slowly moving stable air. These conditions can be expected along cold fronts that have moderate to strong northerly or northwesterly winds behind them. A lull occurs during a frontal passage when the winds to the rear of a weather front are weaker than the winds in advance of the front.

Most of the characteristics of weather fronts that we have discussed thus far apply to all fronts, regardless of whether cold air replaces warm air, or warm air replaces cold air, or whether the front remains in about the same position. So we can label these characteristics as "general" frontal characteristics. But a number of other frontal characteristics must be considered, and these depend upon frontal movements and the stability conditions within the air masses. So, we must classify weather fronts in order to simplify our study of these phenomena. The simplest, and worldwide, scheme of classification is as follows:

> *Cold Front*—a front along which colder air replaces warmer air.
> *Warm Front*—a front along which warmer air replaces colder air.
> *Occluded Front*—a front resulting when a cold front overtakes a warm front and the warm air is forced aloft.
> *Stationary Front*—a front along which one air mass does not replace another air mass.

FRONTAL CLOUDS AND WEATHER

Although temperature (or density) and moisture differences are the primary factors concerning fronts, the cloud systems and weather phenomena associated with the various types of fronts are also of great importance. Warm air, being less dense air, ascends along a frontal surface and is cooled adiabatically (it cools by expansion with no gain or loss of heat). If the warm air is not extremely dry, the water vapor in the air condenses and a cloud system develops along, and mostly above, the frontal surface. If the air underneath the frontal surface is unstable, clouds of secondary importance may also develop below the frontal surface and perhaps even produce precipitation.

Cold Fronts

Of the various types of weather fronts, cold fronts and cold-type occluded fronts usually have the most violent, frontal type of weather associated with them. Figure 7–8 illustrates vertical cross sections of moderately moving or fast-moving cold fronts under various conditions of air-mass stability. By definition, since a cold front is a front along which cold air replaces warm air, the fronts move from left to right in the diagrams. Note that the cloud systems vary with the stability of the air masses. Since the *warm* air *above* the cold frontal surface moves at greater speeds than the *cold* air *below* the frontal surface, the warm air descends along the upper section of the cold front and is heated adiabatically. The heated air becomes relatively dry, and this accounts for the fact that there is usually no cloud system above the upper portion of a cold front. On the other hand, the warm air at lower levels moves more slowly than the cold front and is forced

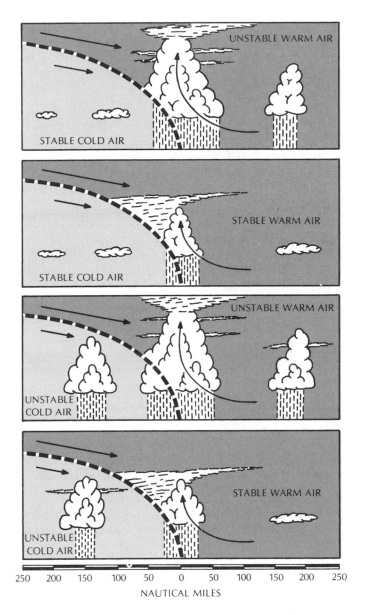

Figure 7–8. Vertical cross sections of moderately or fast-moving cold fronts under various conditions of air mass stability.

upward over the cold frontal surface. This results in the cloud system being above the cold frontal surface and "leaning" in a direction away from the front. If the warm and rising air is stable, the precipitation will be rather "even." If the warm air is unstable, however, squally precipitation, and frequently thundershowers, are superimposed on the other precipitation. If a cold front moves slowly, the descending, warm air currents above the upper portion of the frontal surface are less rapid, and the cloud system extends farther to the rear of the cold front at the ground. In the wintertime, cold fronts move a distance of about 600 miles per day (25 mph). In the summertime, they move more slowly because the pressure gradients are weaker. Table 7–1 is a brief summary of the weather sequence associated with cold fronts.

Squall Lines

Squall lines (sometimes called "instability lines") are extremely dangerous phenomena for small boating operations and should be avoided if at all possible. Squall lines are narrow zones of extreme instability which sometimes occur about 50–300 miles ahead of a fast-moving cold front. These zones of extreme instability are usually oriented parallel to the cold front, and move eastward with about the same speed as the cold front.

The development of squall lines takes place when the winds above the cold frontal surface are moving in the same direction as the cold front, but at a greater speed, and prevent the warm air ahead of the cold front from rising. When this happens, the violent weather (extreme gustiness and torrential downpours) occurs just behind the squall line, rather than at the cold front. From the ground or afloat, an approaching squall line looks like a solid, boiling wall of low, black clouds. Special warnings are broadcast by NOAA's NWS when a squall line threatens an area.

Table 7–1. Summary of Weather Sequence at a Cold Front.

Element	In Advance	During Passage	To the Rear
Weather	Usually some rain; perhaps thunder.	Heavy rain; perhaps thunder & hail.	Heavy rain for short period, then fair. Perhaps scattered showers.
Clouds	Ac or As & Ns, then heavy Cb.	Cb with Fs or scud.	Lifting rapidly, followed by As or Ac; perhaps Cu later.
Winds (northern hemisphere)	Increasing and becoming squally.	Sudden clockwise shift; very squally.	Gusty.
Pressure	Moderate to rapid falls.	Sudden rise.	Rise continues more slowly.
Temperature	Fairly steady; may drop a bit in pre-frontal rain.	Sudden drop.	Continued slow drop.
Visibility	Usually poor.	Temporarily poor, followed by rapid improvement.	Usually very good, except in scattered showers.

Warm Fronts

By our accepted definition, a warm front is a front along which warmer air replaces colder air. Since warm air is less dense (lighter) than cold air, the *cold* air *ahead* of the warm front must be receding. With the warm air being the "lightweight" and the cold air being the "heavyweight" in the tussle, the warm air is in no position to "push" the cold air out of the way.

In the case of warm fronts, the warm air ascending over the wedge of receding cold air produces a gigantic cloud system. The width of this cloud system may extend as much as 600 miles, and the length of the system may be as much as 1,200 miles. The height of the cloud system varies within very wide limits, depending primarily upon the moisture content and distribution, stability, temperature, vertical velocities, and so forth, in the warm air mass. The height of the cloud system may be as little as 8,000 feet or, under proper conditions, as high as 45,000 feet. As a rule, the cloud system extends upward into the layer of sub-freezing temperatures. Because of this, the upper portion of the cloud system will usually contain ice particles. A fairly detailed picture of the clouds and related phenomena associated with a warm front is contained in figure 7–9.

A normal sequence of clouds as a warm front approaches your position is: cirrus, (followed by) cirrostratus, (followed by) altostratus, (followed by) nimbostratus, (followed by) stratus. At this point, it would be a good idea to review the cloud photographs in chapter 3. Because of the great horizontal extent of warm-front cloud systems ahead of the front, the first clues of the impending arrival of a warm front can often be seen two days in advance. When a warm front approaches, an observer at sea (or ashore) will see the clouds arrive in the sequence described above. However, the various cloud types will not be separate and distinct—they will gradually merge into one another. Also, low stratiform or cumuliform clouds (depending on the air-mass stability, and other factors) may be present in the wedge of cold air under the warm frontal surface and obscure some of the upper-level clouds of the warm-front system.

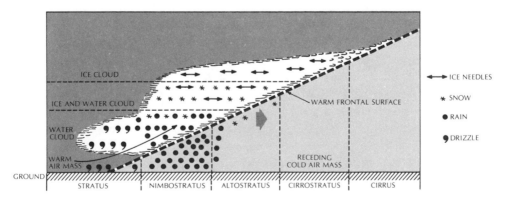

Figure 7–9. Schematic drawing of the clouds and related phenomena associated with a warm front.

If both the cold air mass under the warm frontal surface and the overriding warm air mass are stable, a sort of "smooth" cloud system (as shown in figure 7–9) is the result. Low stratus clouds may, or may not, form in either or both of the warm and cold air masses. When the warm air is stable, the precipitation is steady (or "even"), without sudden changes. When the overriding (ascending) warm air is unstable, however, showery and squall-type precipitation is superimposed on the steady-type, warm frontal precipitation, and warm-front thunderstorms may occur. In the wintertime, warm fronts travel about 360 miles per day (15 mph). In the summertime, like cold fronts, they move more slowly because the pressure gradients are weaker. A brief summary of the weather sequence associated with warm fronts is outlined in table 7–2.

The Rebel Fronts

One of the "puzzlers" for weather forecasters is the so-called *stationary front*. These weather fronts move very little, as the name implies, and their future action is difficult to predict. Sometimes they oscillate back and forth over a ship or station several times. And when there is hot and humid air on one side of the front, and cool, dry air on the other side of the front, these unpredictable oscillations are enough to "blow almost any forecaster's cool." Weather conditions associated with these fronts are somewhat similar to those of warm fronts but are, as a rule, somewhat milder. Because stationary fronts sometimes generate or regenerate quite rapidly into strong fronts, weathermen have to keep close tabs on them.

Table 7–2. Summary of Weather Sequence at a Warm Front.

Element	In Advance	During Passage	To the Rear
Weather	Continuous rain or snow.	Precipitation usually stops.	Sometimes a light drizzle or fine rain.
Clouds	In succession: Ci, Cs, As, Ns; sometimes Cb.	Low Ns and scud.	St or Sc, sometimes Cb.
Winds (northern hemisphere)	Increasing.	Clockwise shift, sometimes decreasing.	Steady direction.
Pressure	Steady fall.	Levels off.	Little change, perhaps slight rise followed by slight fall.
Temperature	Steady or slow rise.	Steady rise, usually not sudden.	Little change or very slow rise.
Visibility	Fairly good except in precipitation.	Poor; often mist or fog.	Fair or poor; mist or fog may persist.

Occluded Fronts

Occluded fronts include both the best and the worst features of cold and warm fronts. And this is to be expected. By our definition, an occluded front is the front resulting when a cold front overtakes a warm front and the warm air is forced aloft. In the process, either the cold front or the warm front is forced to rise off the ground. This depends upon whether the cold air behind the cold front is colder or warmer than the cold air ahead of the warm front. The heaviest (most dense) air mass occupies the bottom position in the triad.

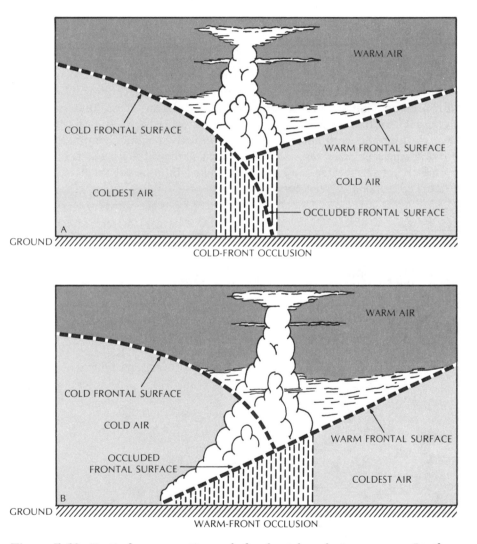

Figure 7–10. Vertical cross sections of the frontal occlusion process. In the upper diagram, the cold air behind the cold front is colder than the cold air ahead of the warm front, and the warm front is forced aloft. In the lower diagram, the cold air ahead of the warm front is colder than the cold air behind the cold front, and the cold front is forced aloft.

Figure 7–10 illustrates cold-front type and warm-front type occluded fronts. In figure 7–10a, the cold air behind the cold front is colder (and heavier) than the cold air ahead of the warm front. The result is a cold-front type occlusion, wherein the warm front is lifted off the ground. Figure 7–10b illustrates the case where the cold air ahead of the warm front is colder than the cold air behind the cold front. The result is a warm-front type occlusion, wherein the cold front is lifted off the ground. In the case of occluded fronts, whether cold-front or warm-front type, precipitation normally occurs on both sides of the front. The cloud and weather sequences of occluded fronts are the combined sequences of cold and warm fronts, as one would also expect.

All Together Now

Figure 7–11 shows how a cold, a warm, and an occluded front and their associated clouds and weather appear on a surface weather map. The isobars have been deliberately omitted from the drawing to reduce the "clutter." Note that cumuliform clouds and showers are typical of the cPk air mass, that stratiform clouds characterize the mTw air mass, and that the steady precipitation is ahead of the warm front and both ahead of, and behind, the occluded front. Note also that in the northeastern part of the mTw air mass (where the air has been cooled by the underlying surface for the longest period of time in its travel), drizzle is falling and (advection) fog has formed.

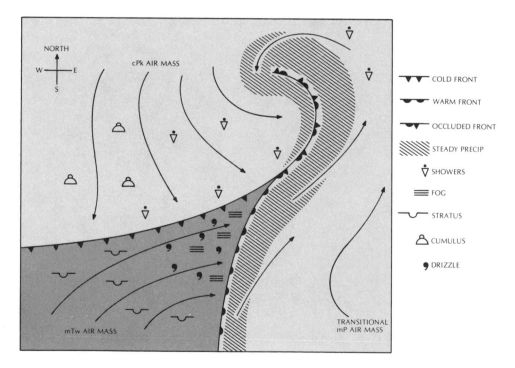

Figure 7–11. Putting it "all together" on a surface weather map. Isobars have been omitted to reduce the "clutter."

THE BIRTH, DEVELOPMENT,
AND DEATH OF A FRONTAL CYCLONE WAVE

The generation of a frontal cyclone wave has still not been fully explained to the complete satisfaction of all atmospheric theoreticians. Nevertheless, there are some statements that can be made in the way of a "descriptive approach" to the subject. Referring to figure 7–12, let's assume that on "day 1" there exists a stationary front in central Texas and southern New Mexico with a cold easterly current to the north of the front and a warm westerly current to the south. The front is stationary because the air currents are parallel to the front. There is no wind component perpendicular to the front.

The frontal surface slopes upward in the direction of the cold air (toward the north) because the cold air is heavier than the warm air. As we ascend upward in the cold air and pass through the frontal surface, the wind changes abruptly from an easterly wind to a westerly wind. There is a strong wind shift (shear) at the frontal surface. Because of this wind shear, waves form on the frontal surface in much the same way as unstable waves form on the ocean surface when there is sufficient wind shear. Through this wave motion, the frontal surface "bulges" up and down. And as it does so, the front at the ground oscillates horizontally as shown on "day 2." If the wave length of the frontal wave is between approximately 400–1,800 miles, the wave is unstable and its amplitude increases with time as shown on "day 3." As the amplitude continues to increase, the cold front overtakes the warm front (remember, cold fronts move faster than warm fronts), and an occluded front is the result, as shown on "day 4." The occlusion process continues and eventually the occluded front dissolves. The cyclone (low-pressure area) develops into a large cyclonic whirl of quite homogeneous air, as shown on "day 5," and in another 24 hours or so, we are back to where we started with a stationary front separating cold air to the north of the front from warm air to the south of the front.

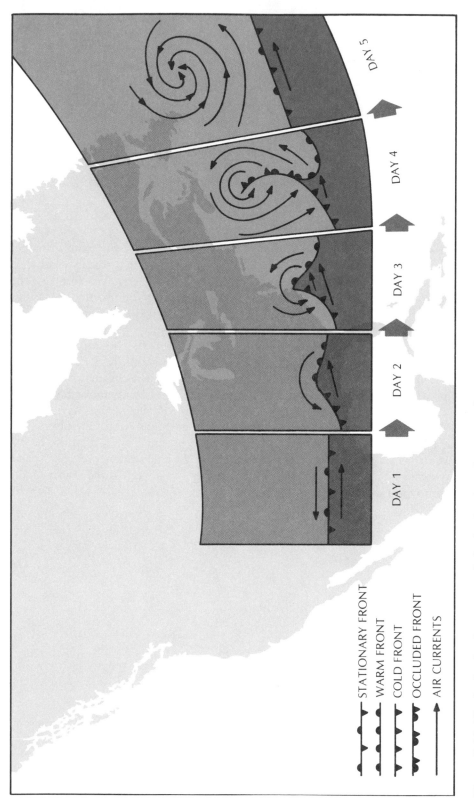

Figure 7-12. The birth, development, and death of a frontal wave cyclone.

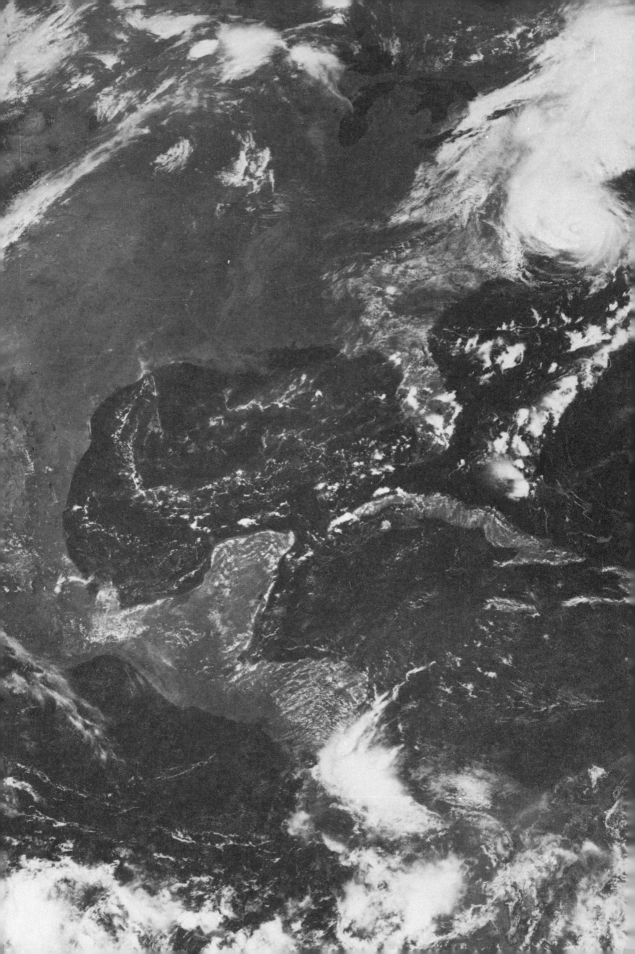

8

weather disturbances and storms

"Big whirls have little whirls
That feed on their velocity,
And little whirls have lesser whirls
And so on to viscosity."

—L. F. Richardson

"Never mind the weather, just so the wind don't blow!" goes an old song that has been sung by mariners for many decades. But the winds *do* blow. They blow comfort when gentle, annoyance when strong, and destruction when violent. In the tropics as well as in middle and high latitudes, the winds swirl over the earth in various patterns—some dangerous, some nonhazardous, and some rather pleasant.

IN THE TROPICS

Compared to the usually rapid—and sometimes violent—weather changes associated with mid- and high-latitude low-pressure systems, the weather in the tropics follows a rather routine schedule most of the time. The frequent seesawing between cold north winds and warm south winds, accompanied by large temperature falls and rises, is missing. Instead, temperature and wind experience a daily cycle which is dictated largely by orographic, coastal, island, or other terrain features.

In some areas of the tropics, particularly in the large trade-wind regions, this dull, daily cycle so dominates the weather that nothing unusual ever seems to occur—except for hurricanes and typhoons. But this view is grossly exaggerated —especially from a weatherman's standpoint. In tropical regions, a cycle of wet and dry seasons replaces the four seasons of middle and higher latitudes, which are determined by temperature. There is a definite sequence of weather changes during the wet (summer) season and sometimes also during the dry (winter) season.

It is a well-known fact that disturbances of one kind or another in the tropics produce more than 90 percent of the rainfall in this region and that a large fraction of the rain falls in a few intense spurts during the year.

There are many different types of tropical disturbances, which vary largely according to their location. Geographical differences are much larger in the tropics than in higher latitudes, where the basic fundamentals of low-pressure systems hold for all areas and all seasons. In the tropics, many unsolved problems remain. The primary reasons for this are the extreme paucity of weather data in tropical latitudes and the lack of funds and personnel for basic and applied research. For our purposes, then, let's limit the discussion to the more well-known types of disturbances in these latitudes.

INTERTROPICAL CONVERGENCE ZONE (ITCZ)

An almost-continuous trough, or belt, of low pressure at the earth's surface extends around our planet in the equatorial regions. This is where the northeast trade winds (of the northern hemisphere) and the southeast trades (of the southern hemisphere) come together—or converge. This region, within which the northeast and southeast trade winds converge, is known by several different names—the Intertropical Convergence Zone (ITCZ), the Equatorial Trough, the Equatorial Front, the Intertropical Front, the Equatorial Convergence Zone, and so forth. In this book, we shall refer to it as the *Intertropical Convergence Zone —the ITCZ.*

Seasonally, the ITCZ migrates from the winter hemisphere into the summer hemisphere, so that the winter hemisphere controls a larger fraction of the oceanic tropics than does the summer hemisphere. Figure 8–1 illustrates the mean positions of the ITCZ during the winter and summer periods of the northern hemisphere. Remember, when it is winter in the northern hemisphere, it is summer in the southern hemisphere, and vice versa.

In general, as shown in figure 8–1, as the ITCZ migrates north and south in the course of a year, it passes the latitudes between its extreme positions twice, lagging behind the sun by about two months. As a result, many parts of the tropics—but by no means all—have two rainy seasons and two dry seasons. Where the ITCZ moves as shown in figure 8–1, the peaks of the rainy seasons occur in December and April on the equator, in November and May at latitude 5 degrees north, and in June and October at latitude 10 degrees north. At latitudes 15 degrees north and 5 degrees south, the two rainy seasons merge into a single broad peak during the summer months.

The total annual meandering of the ITCZ in the western Pacific sometimes amounts to as much as 44 degrees of latitude in extreme cases. As shown in figure 8–1, the ITCZ reaches its northernmost latitude in August and its southernmost latitude in February. It must be realized, however, that the day-to-day variations in the position of the ITCZ may differ substantially from the mean position during both the winter and summer months.

The ITCZ is usually characterized by strong, ascending air currents, a great deal of cloudiness, and frequent heavy showers and thunderstorms. The intensity does, however, vary greatly. Sometimes the ITCZ looks like a tremendous wall of black clouds, with the top extending to 55,000 feet and higher. At other times,

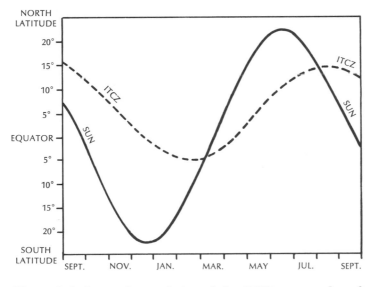

Figure 8–1. Seasonal meandering of the ITCZ compared to the overhead position of the sun. Note the two-month lag of the ITCZ.

it is so weak and inconsequential that it is very difficult to locate with certainty on a weather map, for there is little weather or cloudiness associated with it. The width of the ITCZ varies from about 20 to 150 nautical miles, and as a general rule, the narrower the zone (i.e., the greater the convergence), the more intense is the weather associated with it.

When the ITCZ is located near the equator, only small and weak cyclonic circulations (vortices) can develop within it. But when it migrates away from the equator (at least five degrees or more), the influence of the earth's rotation becomes great enough to transfer sufficient "spin" to the converging air currents to permit tropical cyclones, hurricanes, and typhoons to develop. This is only one reason why aircraft reconnaissance and weather satellites are so very important over tropical oceanic areas where weather data are normally very, very sparse.

EASTERLY WAVES (EWs)

In general, weathermen and other scientists talk about "waves" in a flow current when that current exhibits fairly sinusoidal oscillations. As mentioned previously, and as shown in figure 8-3, one of the most permanent air currents flowing over the earth is that of the northeast and southeast trade winds over the tropical and subtropical oceanic areas of our planet. At times, these easterly trade currents oscillate in a wave-like manner, as shown in figure 8-4, and thus weathermen speak of "waves" in the tropical easterlies, or easterly waves (EWs).

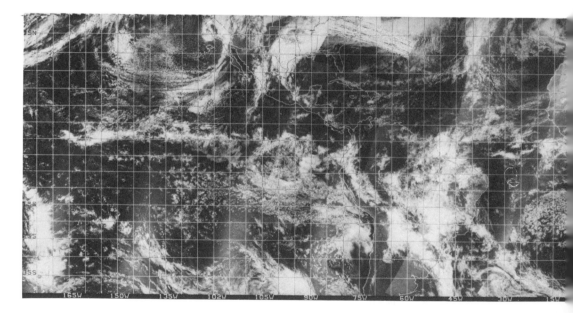

Figure 8–2. The tropical and subtropical regions of the earth as seen by the NOAA-4 polar-orbiting weather satellite on 29 Dec. 1974. Note the great variability of cloudiness associated with the Intertropical Convergence Zone (ITCZ) in the equatorial regions. Courtesy of NOAA National Environmental Satellite Service.

Easterly waves are extremely important phenomena because of their relation to tropical cyclone, hurricane, and typhoon formation. Basically, these waves are troughs of low pressure which are embedded in the deep easterly currents located on the equator side of the large oceanic high-pressure cells centered near 30 to 35 degrees latitude. On the average, easterly waves occur about every 15 degrees of longitude during the summer season primarily and have an average length of about 15 to 18 degrees of latitude. They extend vertically in the atmosphere from the earth's surface to, roughly, 26,000 feet and travel *from east to west* at an average speed of 10 to 13 knots. Note in figure 8–4 that easterly waves are at right angles to the air flow and that the wave amplitude decreases with increasing latitude. Rather than being distinct air-mass boundaries (such as fronts), easterly waves are zones of transition 30 to 100 miles wide, in which the weather changes gradually, but very definitely.

Figure 8–5 is a vertical model of the normal type of easterly wave. Note that the average slope is in a ratio of 1/70 and that the EW slopes upward from west to east. This, coupled with the fact that convergence of air flow occurs to the east of the EW and divergence to the west of the EW, results in the bad weather occurring behind—to the east of—the EW as shown in figures 8–4 and 8–5.

As is true of almost all tropical disturbances, the associated cloud patterns and arrays are best displayed over the open ocean, free from the influences of land surfaces. This is still another reason why aircraft reconnaissance and weather satellites are so very important. Figure 8–6 is a plan-view sketch of the cloud

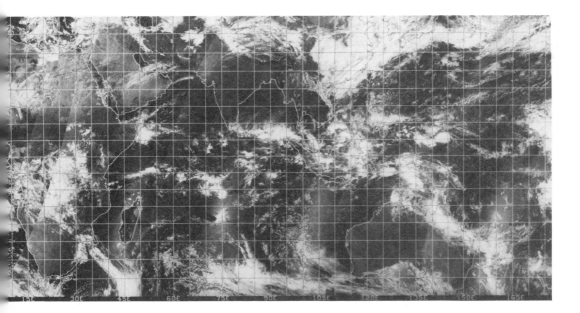

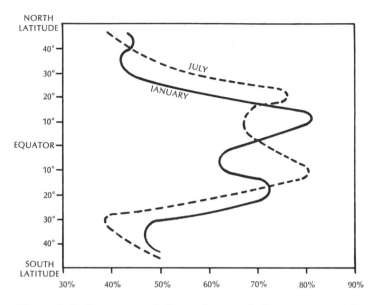

Figure 8–3. Constancy of the surface-wind direction versus latitude expressed in percent. Note that the trade winds attain a constancy of 80 percent.

patterns of a normal easterly wave as would be seen from a high-flying aircraft or a weather satellite. Note that the clouds are arranged almost in rows, forming "cloud streets." To the west of the EW, the rows of clouds are oriented from the northeast to the southwest. The clouds here are small, fair-weather, cumulus clouds, dying out farther west. To the east of the EW, the clouds are oriented from the southeast to the northwest. Heavy cloudiness and bad weather prevail to the east of the EW. A brief summary of easterly wave weather is contained in table 8–1.

MONSOON DEPRESSIONS

Relatively weak cyclones travel over many portions of the oceanic and continental tropics. Surface temperatures are almost constant through these cyclones, or vortices, except for a slight decrease in the area of most intense rain, where evaporation from the falling raindrops cools the atmosphere. Pressure at the center may be two to four millibars lower than in the periphery. Wind speeds average 10 to 20 knots but may increase to 30 to 35 knots close to the center on the north side. As a rule, these cyclones move toward the west or west northwest at 10 to 12 knots.

These vortices are most prominent over southern Asia during the height of the summer monsoon season, when the ITCZ is located on the Asiatic continent, as shown in figure 8–1. These cyclones, or depressions, as they are called in India, move westward, steered by a strong easterly current above 25,000 to 30,000 feet. Except for strong topographic effects, much of the precipitation in Southeast Asia is derived from these cyclones.

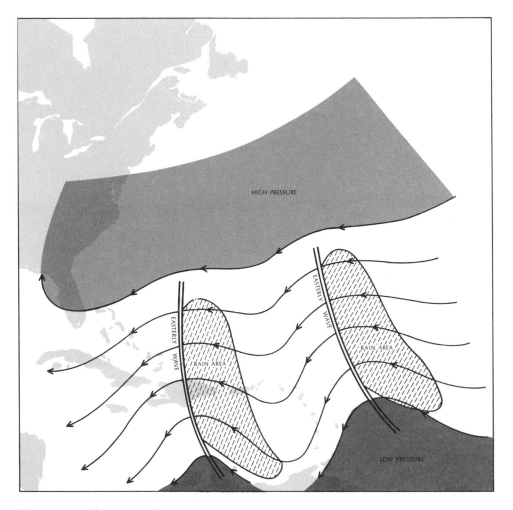

Figure 8–4. Plan view of two easterly waves (EWs) moving westward in the western tropical Atlantic region. The double lines are the EWs. Streamlines show the air flow in the lower levels of the atmosphere.

HURRICANES AND TYPHOONS

From a vantage point in space, hurricanes and typhoons appear as rather small, flat spirals drifting benignly on the sea—gentle eddies in the endless flowing of our planet's atmosphere. Nothing could be more misleading. Where the drift of hurricanes/typhoons (*Hs/Ts*) takes them across shipping lanes and islands and the coasts of continents, their passage is commemorated by the vast destruction of property, the great diminution of prospects—and death.

Hs/Ts are the juvenile delinquents of the tropics, the offspring of ocean and atmosphere, powered by moisture from the sea and the heat of condensation, and driven by the easterly trade winds, the temperate westerly winds, and their own violent energy. In their cloudy and spiraling tentacles and around their

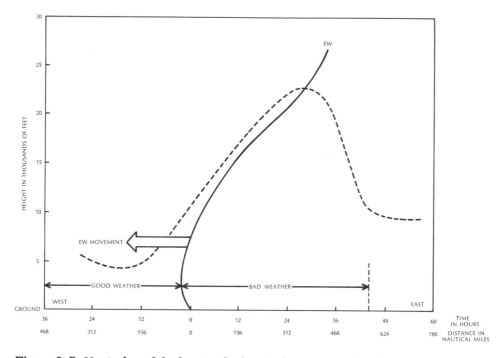

Figure 8–5. Vertical model of a standard easterly wave moving from east to west at 13 knots. The dashed line indicates the depth of the moist layer.

tranquil core, or *eye*, the winds blow with lethal velocity, the ocean develops an annihilative surge, and, as the center moves toward land, tornadoes frequently descend from the advancing wall of thunderclouds.

Compared to the great cyclonic storm systems of middle and high latitudes, Hs/Ts are of rather moderate size, and their worst winds do not attain tornado velocities. Still, their broad spiral base dominates the weather over many thousands of square miles, and they sometimes extend vertically from the earth's surface to over 50,000 feet. See the hurricane model in figure 8–7. H/T winds sometimes exceed 200 knots, and their life-span is measured in days and weeks, not hours. No other atmospheric disturbance combines duration, size, and violence more destructively than Hs/Ts. And over the centuries, seafaring men have watched the sky and ocean with anxiety for signs of an approaching hurricane or typhoon. They feared—and rightly so—encountering the small central area of these tropical cyclones with its phenomenally severe weather. Under the stress of H/T winds, mountainous waves sometimes greater than 50 feet in height are generated in the ocean. Sometimes along concavely curving coasts such as that of the Gulf of Mexico, and in estuaries, the storm surge—the rise of the water above the tide otherwise expected—often exceeds 12 feet, threatening coastal protection works and inhabited areas.

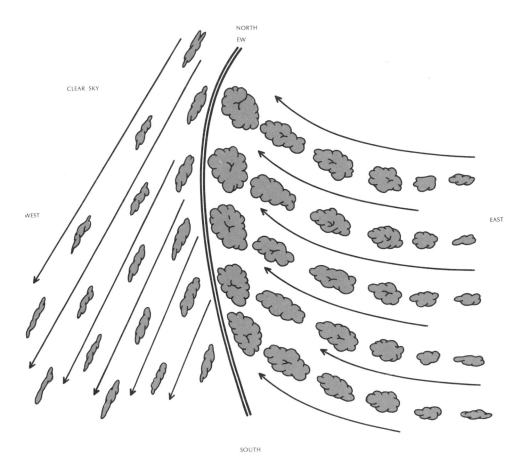

Figure 8–6. Plan view of "cloud streets" associated with a normal easterly wave, as they might be seen from a high-flying reconnaissance aircraft (adapted from J. S. Malkus and H. Riehl).

Description

As mentioned briefly in chapter 1, Hs/Ts are tropical cyclones whose associated wind speeds exceed 64 knots (74 mph). Like all cyclones, they are low-pressure systems—atmospheric pressure systems in which the barometric pressure decreases progressively to a minimum value at the center, and toward which the winds blow spirally inward in a counterclockwise direction in the northern hemisphere. In the southern hemisphere, the winds spiral inward in a clockwise direction.

There are several important differences between tropical cyclones and non-tropical, or "extra-tropical," cyclones. Tropical cyclones are not accompanied by fronts. Extra-tropical cyclones usually are. Tropical cyclones have a warm core, or center (compared to the periphery of the cyclone). Extra-tropical cyclones have a cold core, or center. Tropical cyclones which have developed to hurricane

Table 8-1. Summary of Easterly Wave Weather.

	West of Trough	Close to Trough	At Trough Line	East of Trough
Clouds	Fair-weather cumulus Few build-ups.	Cumulus build-ups. Some high and middle clouds.	Heavy cumulus build-ups. Broken to overcast high and middle clouds.	Heavy cumulus and thunderheads. Layers of high and middle clouds.
Visibility	Strong haze.	Improving.	Good, except in showers.	Fair to poor in showers.
Precipitation	Usually none.	Scattered showers.	Frequent showers.	Heavy showers, thundershowers, and rain.
Surface winds	ENE to NE.	ENE to E.	Easterly, some gusts.	ESE to SSE, with frequent gusts.

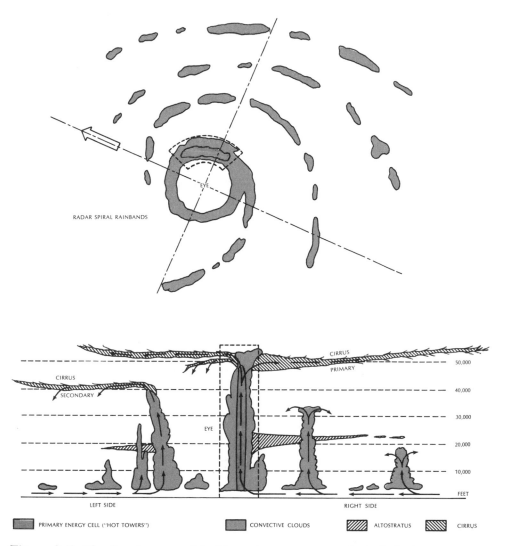

Figure 8–7. The hurricane model. The primary energy cell (called the convective chimney) is located within the area enclosed by the broken line. (From Project *Stormfury* Annual Report, Appendix D, 1965.)

or typhoon intensity often have sustained winds of 120–150 knots. Extra-tropical cyclones almost never do. Tropical cyclones average about 400–500 miles in diameter. Extra-tropical cyclones are much larger, averaging 1,200–1,500 miles in diameter when they are well developed. Tropical cyclones move rather slowly when heading westward and their path of travel is similar to a sine curve, rather than a straight line. After recurving to a northeasterly direction of movement, tropical cyclones accelerate rapidly, sometimes reaching speeds of translation of 50 knots (58 mph). Extra-tropical cyclones seldom travel toward the west.

Whereas extra-tropical cyclones have an almost infinite variety of shapes, usually elongated, tropical cyclones are amazingly symmetrical—almost circles. Tropical cyclones develop in the warm and moist E and mT air masses. Extra-tropical cyclones usually develop along fronts where two different air masses are in "juxtaposition." Tropical cyclones disintegrate fairly rapidly when they move over a land surface and their supply of moisture is cut off. This is their primary energy source. They also disintegrate, or are transformed into an extra-tropical system, when there is an injection of very cold air into the center, or when they travel over a cold water surface. At the center of hurricanes and typhoons is found an almost-calm and almost-cloudless "eye." Extra-tropical cyclones do not have an "eye." The strongest winds associated with Hs/Ts are located near the center. The strongest winds associated with extra-tropical cyclones are usually found in the periphery of the storm.

Formation

Many scientific papers have been published on the subject of hurricane and typhoon (H/T) formation, but there is still no universally accepted theory. We have the "convectional hypothesis," the "waves in a baroclinic easterly current" theory, the "dynamic instability" theory, and several others. However, each theory deals with only one or two specific aspects of tropical cyclone formation, and there is no composite treatment of all aspects. Excellent work in this field is being accomplished at NOAA's NWS National Hurricane Center in Miami, at its Environmental Research Laboratory, and at several universities. Encouraging progress is also being made in the computer modeling of hurricanes, and one day we shall have the answers. But an exact theory is not necessary for our purposes.

Severe tropical cyclones (hurricanes, typhoons, and all the other names for the same phenomena) occur in all warm-water oceans *except* in the South Atlantic Ocean. This is because the ITCZ never invades the South Atlantic. In order for a cyclonic circulation to develop, the converging air currents must be able to concentrate their *vorticity* (local rotation, or "spin," in a fluid flow) at any altitude. The vorticity is very small near the equator and zero on the equator. As a result, Hs/Ts almost never form within 5° north or south of the equator. The absorption of heat from the ocean by the air is greater when a circulation develops in the areas and during the seasons with the highest ocean-surface temperature. This occurs over the *western* parts of the oceans during August–October in the northern hemisphere, and during January–March in the southern hemisphere.

One situation that is very conducive to the formation of a hurricane or typhoon (if other conditions are favorable) occurs when an easterly wave (EW) moves into juxtaposition with a northward "bulge" of the ITCZ, as shown in figure 8–8. The NE trade winds of the northern hemisphere have already developed a cyclonic pattern because of the presence of the EW. As the SE trade winds of the southern hemisphere (where winds are deflected to the left) cross the equator and enter the northern hemisphere, they are then deflected to the right and become SW winds, just about completing a closed cyclonic circulation.

Hurricanes and typhoons derive their tremendous energy (and violence) from the latent heat of condensation (598 calories per gram) released into the atmosphere as the water vapor condenses. As long as the cyclonic center remains

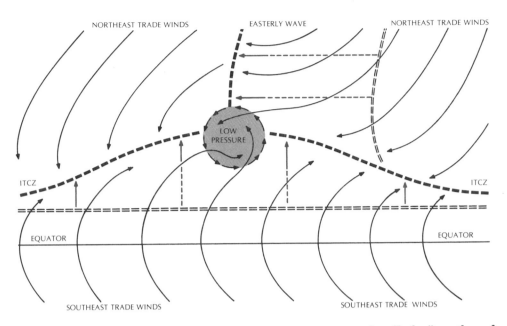

Figure 8–8. When an easterly wave moves into juxtaposition with a "bulge" northward of the intertropical convergence zone (ITCZ), the surface air currents are favorable for the formation of a hurricane or typhoon. The dashed lines indicate the previous positions of the EW and the ITCZ. Dashed arrows show the movement.

over warm water, the supply of energy is almost limitless. As more and more moist E or mT air spirals inward toward the low-pressure center to replace the heated and rapidly ascending air, more and more heat is released into the atmosphere, and the wind circulation continues to increase. When the wind speeds exceed 64 knots (74 mph), we have, by definition, a hurricane or typhoon. Although we have no universally accepted, detailed theory of formation, most meteorologists agree that certain conditions are favorable (or necessary) for hurricane and typhoon formation. These conditions are contained in table 8–2.

Table 8–2. Conditions Favorable (or necessary) for The Formation of Hurricanes and Typhoons.

1. Sea-surface temperature higher than 26°C (78.8°F).
2. Below-normal pressure in low latitudes (less than 1004 millibars, or 29.65 inches) and above-normal pressure in higher latitudes.
3. An existing tropical disturbance of some sort at the earth's surface.
4. Movement of the disturbance at a speed less than 13 knots.
5. Easterly winds decreasing in speed with height, but extending upward to at least 30,000 feet.
6. Special dynamic conditions in the air flow near 40,000 feet.
7. Heavy rain or rainshowers in the area.

Source Regions and Tracks

As mentioned previously, severe tropical storms occur in all of the tropical oceans except the South Atlantic. And the southwestern portion of the North Pacific has more severe tropical cyclones (typhoons) than any other place on earth. Because of the large newspaper headlines given to the devastation and havoc wreaked by these beasts of nature, it may seem surprising that less than 10 percent of tropical disturbances develop into full-blown hurricanes or typhoons. Even in a very active year in the tropics, seldom will there be more than 50 severe tropical cyclones per annum reaching hurricane force in the northern hemisphere. Compare this number to the average of about 2,500 extra-tropical cyclones in the winter months in the northern hemisphere. Also, the number of severe tropical cyclones varies within wide limits from year to year. In the North Atlantic area, for example, the number has ranged from zero to 12 during the hurricane season. Taken over a long period of time, table 8–3 shows the *average* number of severe tropical cyclones (surface winds in excess of 64 knots, or 74 mph) per year for the various geographical areas throughout the world.

Forecasting the future tracks and speeds of movement of hurricanes and typhoons is one of the most challenging and difficult tasks that any operational meteorologist has to face. And each forecaster is well aware of the "sword of Damocles" hanging over his head because so much hinges on his hurricane/typhoon forecasts—the evacuation of many thousands of people from coastal and low-lying areas, the expenditure of millions of dollars in boarding up properties and moving ships and aircraft, the possibility of ships and boats being sunk or

Table 8–3. Average Number of Severe Tropical Cyclones per Year (Winds Greater Than 64 Knots) by Geographical Area.

North Atlantic Ocean (Including Gulf of Mexico and the Caribbean)	8
North Pacific Ocean (Off the SW coast of U. S. and W coast of Mexico)	7
North Pacific Ocean (W of 180th Meridian)	24
North Indian Ocean (Arabian Sea)	2
North Indian Ocean (Bay of Bengal)	6
South Indian Ocean (NW Australia)	2
South Indian Ocean (W of 90°E)	7
SW Pacific and E of Australia	2
Off the SE coast of Africa	2

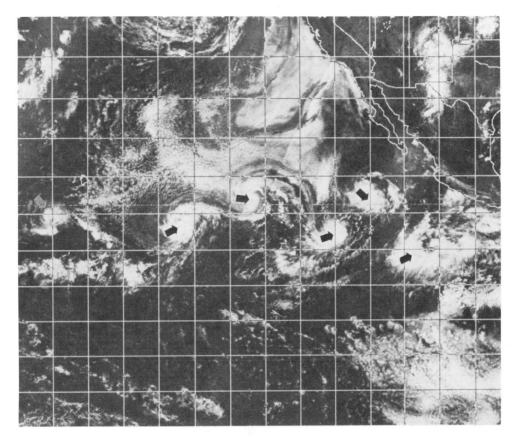

Figure 8–9. Six severe tropical cyclones at one time (in various stages of formation, development, and decay) in the eastern Pacific, as seen by one of NOAA's weather satellites. August 24, 1974. Courtesy of NOAA National Environmental Satellite Service.

ending up "high and dry," the potential of the large loss of human life, and so forth. In addition to his vital forecasts, a hurricane meteorologist has many, many things on his mind.

As a result of aircraft reconnaissance, land station, shipboard, and aircraft weather and radar reports, weather satellites, and the accumulation of large quantities of weather and oceanographic data through the years, we know that simple solutions to hurricane and typhoon movement seldom provide the correct answers. So-called *unusual* tracks occur far too often. Too many tracks exhibit humps, loops, staggering motions, abrupt course and/or speed changes, and so forth.

No two, recorded, severe, tropical, cyclone tracks have ever been *exactly* the same in any ocean. However, an examination of many hundreds of tracks reveals that there are 12 different basic tracks that severe tropical cyclones tend to follow. These are shown in figure 8–10. It must be remembered, however, that even these 12 basic tracks have many minor variations.

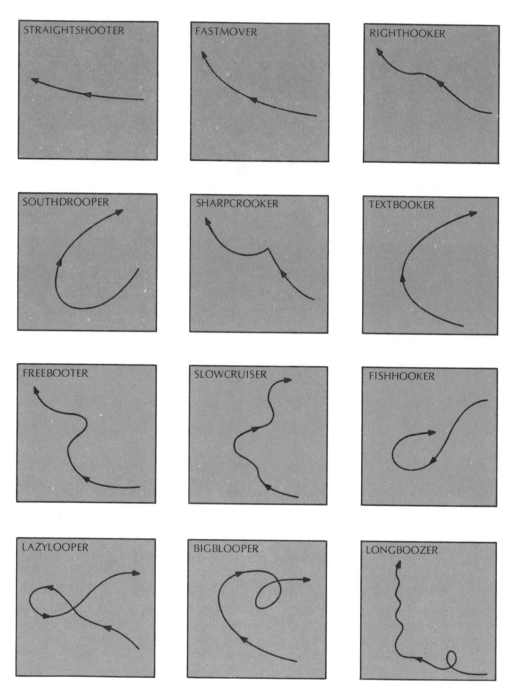

Figure 8–10. The 12 basic tracks of severe tropical cyclones. (From *Heavy Weather Guide* U. S. Naval Institute, 1965).

Hurricanes and typhoons tend to move under the influence of both internal and external forces or influences. Internal forces arise within the circulation of the H/T itself. External forces are applied by the air currents which surround the H/T on all sides and carry it along. Despite all of the unusual facets and irregularities of movement illustrated in figure 8–10, all severe tropical cyclones have one thing in common: a tendency to move eventually toward higher latitudes. This tendency, shared by cyclones at all latitudes (and in both hemispheres), suggests propulsion by an internal force. Additionally, it can be demonstrated mathematically that severe tropical cyclones will move in the general direction of the "steering air current" in which they are embedded and with a certain percentage of the current's speed. A *steering current* is defined as the pressure-weighted, mean air flow from the earth's surface upward to about 30,000 feet (a pressure of about 300 millibars) and over a latitudinal band width of about 8° and centered over the severe tropical cyclone.

In the language of most weathermen, the term *recurvature* in tropical cyclone work refers to the change in the direction of movement from a westerly to an easterly component of movement. That is, a hurricane or typhoon *curves* from a course of 270° to 340° (it is still heading toward the west). But it *recurves* from a course of 340° to 030°. (When the direction of movement passes 360° and continues turning toward the right, it is heading toward the east.) Forecasting the exact recurvature of tropical cyclones in either hemisphere is one of the most difficult problems confronting meteorologists.

Prior to recurvature, the small-scale track of tropical cyclones is usually close to being sinusoidal, as shown in figure 8–12, leading some forecasters astray in their predictions. The speed of forward movement at this stage is quite slow (4–13 knots). After recurvature, when the tropical cyclone has moved into higher latitudes, the track is reasonably "straightline" and forward translational speeds of 40–50 knots are not uncommon.

The Eye

The *eye* of a hurricane or typhoon, as seen in figure 8–13, is unique; it is not observed in any other weather phenomena. It has always been one of the favorite subjects of writers and investigators, for in some respects, it is the most spectacular part of the hurricane's anatomy.

As we discussed earlier in hurricanes and typhoons, the winds spiral violently inward toward the center of lowest pressure. There, the strongly converging air is whirled upward by convection, by the mechanical thrusting of other converging air, and by the pumping action of high-altitude circulations. This spiral is marked by the thick wall clouds which surround—and form the outer edge of—the hurricane's center. The wall cloud of a H/T is the area of most violent winds, heaviest precipitation, and greatest release of heat energy. This ring of strongest winds and torrential rain is usually 5 to 30 miles from the H/T's center, with the average eye diameter being about 15 nautical miles. Inside the eye wall cloud, the winds decrease rather abruptly to 12 knots or less and the torrential rain ceases. This is the celebrated H/T *eye*—a term coined many decades before the advent of radar or weather satellite pictures. The eye offers a brief, *but very deceptive*, respite from the extreme weather conditions of the eye wall. Inside the eye, the winds

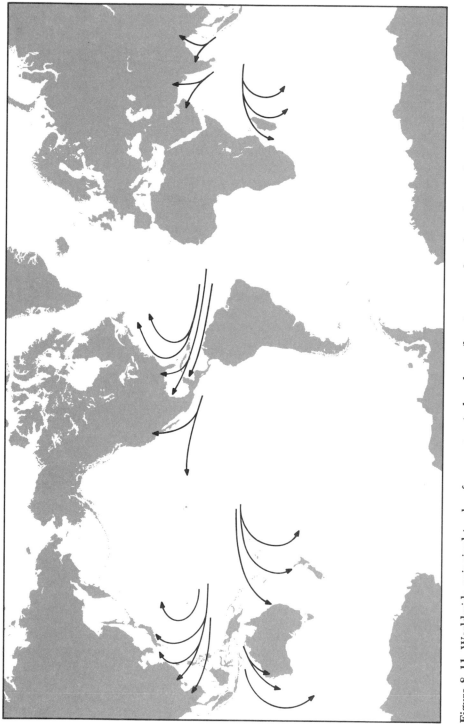

Figure 8–11. Worldwide principal tracks of severe tropical cyclones (hurricanes and typhoons), greatly simplified. Note that there are no tracks in the South Atlantic Ocean. Why?

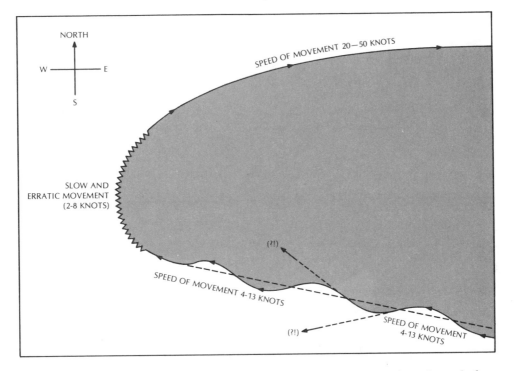

Figure 8–12. Speed, direction, and nature of movement of tropical cyclones before, during, and after recurvature. Note sinusoidal nature of track when the cyclone is heading toward the west. This sometimes misleads forecasters.

are light and there are intermittent bursts of blue sky and sunlight through the thin middle and high clouds. Occasionally, low clouds form miniature spirals inside the eye. Across the eye, at the opposite wall, however, the torrential rain and violent winds resume—but come from the *opposite* direction because of the cyclonic circulation of the storm.

To primitive man, this fantastic transformation from storm violence to comparative calm, and from calm into violence again from another quarter—all in a relatively short interval of time—must have seemed an excessive whim of the gods. Still, it *is* spectacular. The eye's relatively calm and abrupt existence in the midst of such meteorological violence is not soon forgotten by anyone experiencing this phenomenon.

Although the eye of a hurricane or typhoon is usually described as circular, it is frequently observed by aircraft reconnaissance crews to be elliptical. When this is the case, the longest axis is usually parallel to the direction of the storm's movement. Aircraft crews have also observed that the diameter of the eye at 10,000 feet is just about twice the diameter at the earth's surface, and that the eye is constantly undergoing transformation. This is hardly surprising, however, when one considers the extreme conditions at the boundary of the eye.

Figure 8–13. High-altitude view from an aircraft showing a hurricane's eye, wall clouds, and spiral rain bands. Official U. S. Navy Photograph.

Figures 8–14 to 8–18 show a superb series of radar photos of a Sugar Mike 10-centimeter radarscope. These photos are a dramatic description of the U.S. Third Fleet's devastating encounter with the eye of a typhoon in the western North Pacific in 1944. During this typhoon three destroyers capsized and went down, nine other ships sustained serious damage, and 19 other vessels received lesser damage. The magnitude of the disaster concerned Admiral C. W. Nimitz, USN (then Commander in Chief, U.S. Pacific Fleet), who wrote in part: "The time for taking all measures for a ship's safety is while still able to do so. Nothing is more dangerous than for a seaman to be grudging in taking precautions lest they turn out to have been unnecessary. Safety at sea for a thousand years has depended on exactly the opposite philosophy."

Pressure and Wind

As we discussed in chapter 5, motion in the earth's atmosphere—the wind—is a function of the pressure difference between two points. In general, the winds tend to blow across the isobars at a certain angle, from higher toward lower pressure. The greater the pressure difference, the faster the wind blows. Obviously, then, the violent winds, such as those found near the centers of Hs/Ts, are indicative of extremely large pressure differences. The pressure at the centers

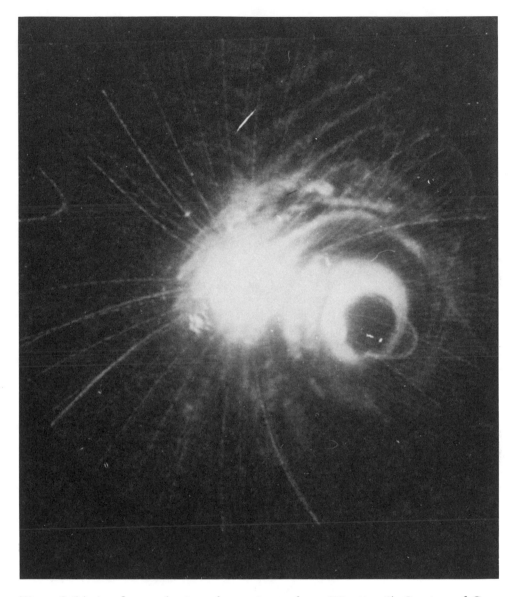

Figure 8–14. A radarscope's view of a passing typhoon (Situation 1). Courtesy of Captain Edwin F. Barker, Jr., USN.

Time	1100 Local Time
True Bearing and Distance of Eye	077°; 39 Nautical Miles
Wind	336°; 57 Knots; Gusts to 66 Knots
Pressure	994.5 mbs (29.37 inches)
Ceiling	Less than 500 Feet
Visibility	800 to 1,200 Yards
Waves/Swell	Very High (to 40 Feet)

Figure 8–15. A radarscope's view of a passing typhoon (Situation 2). Courtesy of Captain Edwin F. Barker, Jr., USN.

Time	1130 Local Time
True Bearing and Distance of Eye	068°; 38 Nautical Miles
Wind	323°; 68 Knots; Gusts Over 80 Knots
Pressure	995.0 mbs (29.38 inches)
Ceiling	Less than 500 Feet
Visibility	800 to 1,200 Yards
Waves/Swell	Very High (20 to 40 Feet)

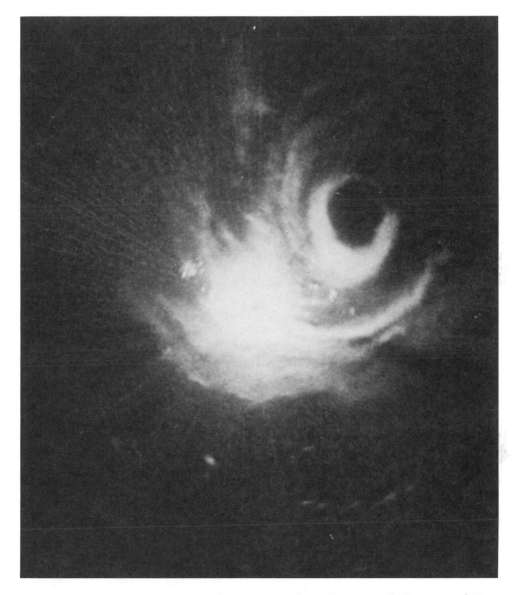

Figure 8–16. A radarscope's view of a passing typhoon (Situation 3). Courtesy of Captain Edwin F. Barker, Jr., USN.

Time	1200 Local Time
True Bearing and Distance of Eye	055°; 40 Nautical Miles
Wind	315°; 65 Knots; Gusts Over 75 Knots
Pressure	996.1 mbs (29.42 inches)
Ceiling	Less than 500 Feet
Visibility	600 to 1,000 Yards
Waves/Swell	Very High to Mountainous (in excess of 40 feet)

Figure 8–17. A radarscope's view of a passing typhoon (Situation 4). Courtesy of Captain Edwin F. Barker, Jr., USN.

Time	1340 Local Time
True Bearing and Distance of Eye	005°; 38 Nautical Miles
Wind	270°; 49 Knots; Gusts Over 75 Knots
Pressure	993.2 mbs (29.33 inches)
Ceiling	Less than 500 Feet
Visibility	400 to 800 Yards
Waves/Swell	Very High to Mountainous (in excess of 40 feet)

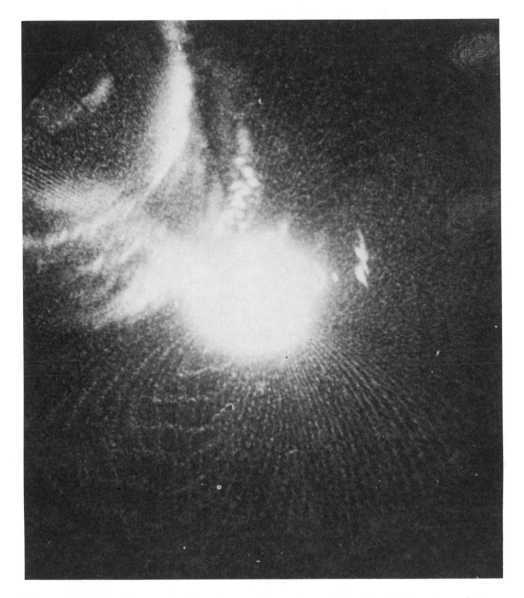

Figure 8–18. A radarscope's view of a passing typhoon (Situation 5). Courtesy of Captain Edwin F. Barker, Jr., USN.

Time	1530 Local Time
True Bearing and Distance of Eye	326°; 81 Nautical Miles
Wind	250°; 41 Knots; Gusts Over 60 Knots
Pressure	1000.1 mbs (29.53 inches)
Ceiling	7,000 to 8,000 Feet
Visibility	6,000 Yards Off Bow, Less than 1,000 Yards Off Stern
Waves/Swell	High to Very High (12 to 20 feet)

of Hs/Ts is very low in order that the rapid decrease in pressure in a short horizontal distance be established and maintained to support the devastating winds.
In a general way, the pressure in the center of a H/T is a measure of its intensity;
the lower the pressure in the center, the more intense the H/T. To the writer's
knowledge, the lowest scientifically accepted sea-level barometer reading occurred in a typhoon 460 miles east of Luzon—886.56 millibars (26.18 inches).

Figure 8–19 is a typical example of a barograph trace as a H/T approaches,
passes over, and leaves a ship's position or station. On the day before the H/T's
arrival, local weather is often unusually good with the barometer reading above
normal. Then, the pressure begins to fall slowly (in excess of the diurnal value)
and the wind may start blowing from an unusual direction. Finally, the pressure
falls at an extremely rapid rate (usually for about three hours) as the eye, itself,
approaches, and rises almost equally rapidly as the eye moves away. Pressure falls
of as much as 40 millibars (1.181 inches) in 20 minutes have been recorded, and
total pressure drops of 60 millibars (1.772 inches) in 50 miles horizontally are
not uncommon.

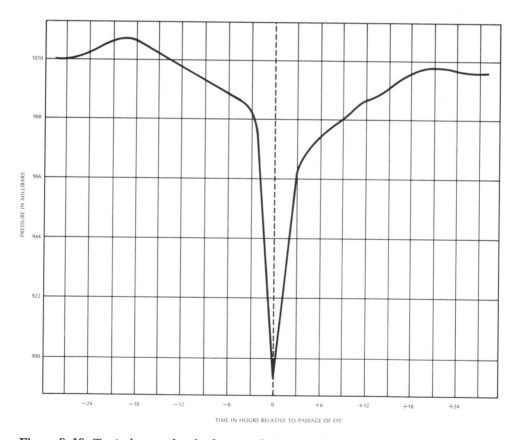

Figure 8–19. Typical example of a barograph (pressure) trace as the eye of a hurricane
or typhoon approaches and passes over a ship or station. (From *Heavy Weather Guide*,
U.S. Naval Institute, 1965).

The highest H/T winds have never been scientifically recorded simply because the recording instruments have collapsed and have been carried away long before the greatest intensity of the storm is reached. Maximum estimates are based generally on: (1) aircraft reconnaissance flights, (2) calculations from pressure-gradient measurements and pressure-temperature relationships, and (3) detailed studies of structural damage. The horizontal distribution of wind speeds around a typical H/T is illustrated in figure 8–20. This diagram shows a maximum of 150 knots (173 mph) to the right of the direction of movement of the center, looking downstream. It is characteristic of a moving H/T that the strongest winds occur to the right of the center. In the case of stationary centers, the distribution of wind speeds is much more symmetrical. Figure 8–21 illustrates the vertical distribution of wind speeds around a H/T center.

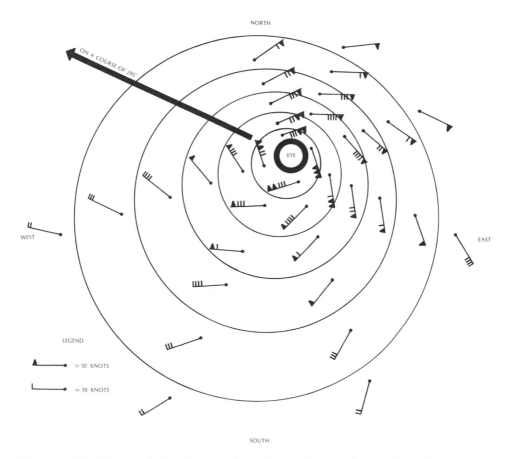

Figure 8–20. Horizontal distribution of wind speeds around a 150-knot hurricane or typhoon in the northern hemisphere. Each pennant signifies 50 knots of wind and each barb indicates 10 knots. Thus, two pennants and two barbs equal 120 knots, and one pennant and two barbs equal 70 knots. (From *Heavy Weather Guide*, U.S. Naval Institute, 1965).

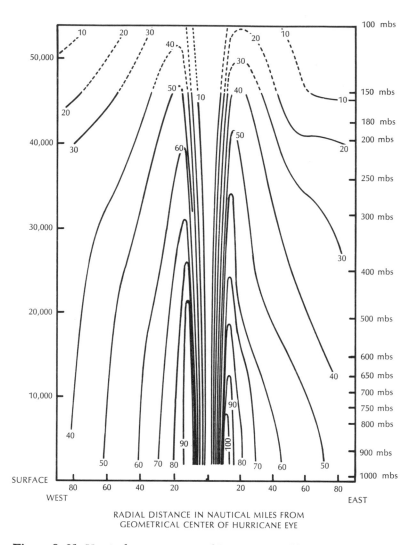

Figure 8–21. Vertical cross section of Hurricane Hilda (1964) showing the relative wind speeds in knots. Courtesy of Hawkins and Rubsam.

Evidence is available from gust recordings which indicates that the extreme gusts frequently exceed the maximum sustained winds by as much as 30 to 50 percent. This means that for a 150-knot H/T, one should expect extreme gusts as high as 225 knots.

One must remember that the force exerted by the wind does not increase proportionately with velocity, but rather with the square of the velocity. Doubling the wind velocity results in four times the force! Thus, a 52-knot wind results in a force of 15 pounds per square foot, but a 110-knot wind results in the terrific force of 78 pounds per square foot. The problem is quite complicated, however. H/T-force winds passing over and around ships and installations develop both positive and negative forces. Positive forces are those pushing in, and negative forces are those pulling out (suction effect) externally. For certain angles of incidence of the wind, the negative pressure may be considerably larger than the frontal force. The gustiness of the wind results in the uneven, intermittent pressures and wrenching effects that cause so much damage—especially to tall buildings and towers.

Clouds and Precipitation

Even though the minute details of cloud sequence vary from one H/T to another, the precursory signs are remarkably uniform. On the day before the H/T's arrival, local weather is frequently unusually good. There are only a few cumulus-type clouds, and these do not extend very high. Late in the afternoon on the day before the H/T's arrival, high-level cirrus clouds approach from the direction of the center. One begins to feel muggy. The cirrus clouds are followed in order several hours later by lowering and thickening cirrostratus, altostratus, and altocumulus clouds. Then, several brief periods of tall cumulus clouds with increasing shower activity are experienced. The showers become much heavier, more frequent, and the wind begins to increase substantially. Finally, a dark wall of clouds approaches—the *bar* of the storm. With its arrival, the full fury of the H/T is unleashed.

This fury is not constant. Periods of extreme violence alternate with periods of much less intensity. Torrential rain is interspaced with periods of only moderate rain. Heaviest rain occurs under the wall cloud and under the spiral cloud bands, whose shape is close to that of a logarithmic spiral. Some of the world's heaviest rainfalls have occurred in connection with Hs/Ts. In 1911, one typhoon inundated the Philippines with 88 inches of rain in a four-day period—which is almost the total annual rainfall on the island of Guam. Exact rainfall of a H/T is almost never accurately known. After the wind speed exceeds 50 knots or so, it is unlikely that the rain gauges collect more than 50 percent of the rainfall. And when the winds go higher, the rain gauges become a part of the lethal, flying debris. Total rainfall at a particular locality is, of course, largely dependent upon the speed of movement of the H/T, simply because in slowly moving storms the rain lasts much longer.

Waves, Swells, and Floods

Waves are generated by the wind, and as these wind-generated waves move out of, or ahead of, the generation area and into regions of weaker winds, they decrease in height and increase in period. These decaying waves, which persist for a long time, are termed *swells*.

The great wind stress exerted on the surface of the sea in Hs/Ts produces huge waves with phenomenal heights. Some have been reported as high as 66 feet. Figure 8–22 shows schematically how these waves travel outward in all directions from the center of the H/T. These waves are observed as a long swell as much as 2,000 miles distant from the storm center. The speed of propagation of these long waves is as high as 1,000 miles per day. Since the average H/T speed of movement prior to recurvature is only 330 miles per day, these swells provide an excellent warning of an impending H/T.

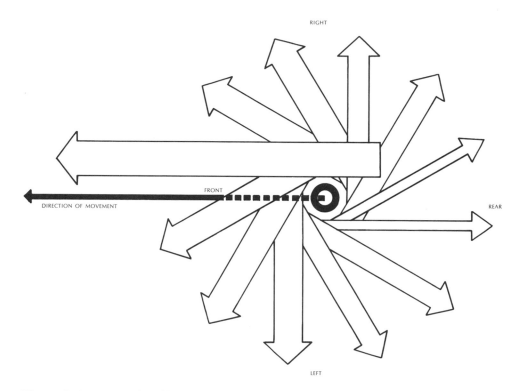

Figure 8–22. Waves/swells generated by a hurricane or typhoon. Arrows indicate the direction of movement of the waves/swells. The width of the arrows indicates relative height. (*Heavy Weather Guide*, U.S. Naval Institute, 1965).

Referring again to figure 8–22, the waves generated in the right rear quadrant of the storm travel in the direction of the storm's movement. Thus, they propagate under the influence of winds having relatively little change in direction for a much longer time than the waves in any other quadrant. Consequently, the highest waves (and later, swells) are produced in this part of the storm. When these swells arrive at a distant ship (or coastline or island), the normal wave frequency of 10 to 15 per minute will have decreased to about 2 to 5 per minute. The direction of the swell indicates the position of the H/T's center at the time the swells were generated.

If the swell direction remains constant, the H/T is approaching the ship (or area) directly. If the swell changes counterclockwise, as seen by an observer facing the storm, the center of the H/T will pass from right to left. If the swell changes clockwise, the center of the H/T will pass from left to right.

One must remember that a cubic yard of water weighs three-fourths of a ton, and a breaking wave moving toward shore at speeds of more than 50 knots is one of the most destructive elements connected with a hurricane or typhoon. The erosive power of these storm-driven waves is tremendous, for these waves can scour out hundreds of feet of beach in just a few hours.

The *storm surge* occurs just to the right of the H/T's center, and either shortly precedes or accompanies the arrival of the center. When the surge moves into a long channel, however, the rise in the water at the end of the channel may reach its peak several hours after the H/T winds have passed. A H/T moving toward or crossing a coastline will always be accompanied by above-normal tides, particularly near and to the right of the center. These high tides frequently occur at a considerable distance ahead of the advancing center, providing another excellent warning sign.

Inundation by torrential rains, storm surges, and floods has been responsible for the deaths of millions of people and property damage in the billions of dollars through the years. No area affected by hurricanes and typhoons has been immune. But the extent of loss of life and destruction of property has been largely dependent upon the physical characteristics of the drainage basin, the rate and total accumulation of rainfall, and the river stages at the time the rains begin.

Rules for Precaution or Disengagement

The following precautionary rules are offered:
1. If the surface pressure drops to 1002 millibars (29.59 inches) in your location, start thinking about preliminary buttoning-up operations for your craft, whether or not you have received a "warning." If the surface pressure drops below 1,000 millibars (29.53 inches), start the preliminary buttoning-up. Your receipt of the warning may have been delayed for some reason.
2. If you notice an unusual long swell with a frequency of about two to five crests per minute, the chances are that there is a storm center at some distance from your position, in the general direction from which the swell is approaching.
3. If the center is anywhere near your position, keep a detailed plot of the six-hourly warnings which are broadcast.
4. Never try to outrun or pass ahead of a H/T center if there is some other course of action available to you. Chances are that you will *not* make it. Remember that the H/T-generated swells move ahead of the storm at speeds close to 50 knots. Once these swells arrive at your vessel, your speed of advance is cut down to four or five knots, and the eye could pass right over you. If you should become caught in the eye, don't let the relative calm fool you. Remember that after the calm, the wind will blow even more furiously from the opposite direction.
5. The winds are the best indication of the direction of the center from your craft. Memorize the diagram of figure 8–20. Adding 115 degrees to the direction from which your observed true wind is blowing gives the approximate bearing from your craft to the H/T center. For example, if your craft bears 090 degrees from the center, the observed true wind would be from 155 degrees. Adding 115 degrees to 155 degrees, the true bearing of the eye from your craft would be 270 degrees. For sums greater than 360 degrees, subtract 360 to obtain the bearing.
6. If circumstances preclude avoiding the H/T altogether, the proper action to take depends upon your position relative to the H/T's center and its direction of travel. If possible, avoid the dangerous semicircle (the right semicircle,

looking downstream, in the northern hemisphere). Try to maneuver into the navigable semicircle (the left semicircle, looking downstream, in the northern hemisphere). The reverse is true in the southern hemisphere.

7. A gradually veering (changing clockwise) wind indicates that you are in the dangerous semicircle. A backing (changing counterclockwise) wind indicates that you are in the navigable semicircle.

8. If the wind direction remains steady with increasing speed and a rapidly falling barometer, you are near (or in) the path of the H/T. If the wind direction remains steady with decreasing speed and a rising pressure, you are on the H/T's path, but in a safe position behind the center.

9. If you are fortunate enough to have radar equipment, be sure to keep it "fired up." This will give you the bearing and distance of the eye—as long as the equipment holds up.

10. Do not guess. Work out your relative movement plot and the closest point of approach very carefully on a maneuvering board or your radarscope.

11. In general, if unavoidably caught within the circulation of a hurricane or typhoon—other things being equal—the best courses of action are as follows:

    In the right or dangerous semicircle bring the wind on the starboard bow (045 degrees relative), hold course and make as much speed as possible.

    In the left or navigable semicircle bring the wind on the starboard quarter (130 degrees relative), hold course and make as much speed as possible.

    On the storm track—ahead of the center—bring the wind two points on the starboard quarter (158 degrees relative), hold the course, and make as much speed as possible. When well within the navigable semicircle, bring the wind to 130 degrees relative.

    On the storm track—behind the center—steer the best riding course that maintains a constant or increasing distance between your position and that of the H/T. Remember, however, that Hs/Ts tend to curve northward and recurve eastward.

12. Hopefully, you will have safely moored your craft or evacuated it to a safe haven long before the storm's arrival. Once this is accomplished, do not return to the scene until all danger has passed. Remember that weather forecasters are not wizards or magicians. They, too, make mistakes, and allowances should be made for errors in forecasts.

Hurricane Modification by Man

Experiments in modification of hurricanes are a natural outgrowth of other research efforts and are based on facts learned about these storms since the NOAA's NWS National Hurricane Research Laboratory was formed in 1959. *Project Stormfury* is a joint Department of Commerce/Department of Defense program of scientific experiments designed to explore the structure and dynamics of hurricanes/typhoons and tropical storms. Participating agencies are the U.S. Navy and the U.S. Air Force from the Department of Defense, and NOAA from the Department of Commerce. The Project's objectives are to achieve a better understanding, improve prediction, and examine the possibility of modifying some aspects of these storms.

Hurricane damage is caused by wind, rain, and flood, but principally by the wind-driven storm surge which sends the sea onto the land. If the wind speed, and consequently the wind force, of hurricanes can be decreased as the hurricanes approach land, both death and damage can be reduced to a large degree.

Since the beginning of the 20th century, the hurricane death toll has declined markedly, as observation, prediction, and preparedness have improved. In the decade 1900 through 1909, more than 8,000 people were killed by hurricanes in the United States alone. Since 1940, hurricane-caused deaths exceeded 500 in only three cases. But the decreasing loss of life is "mirrored" by the sharply rising property damage during the same period. Adjusting damage totals to the 1957–59 base of the Commerce Department's composite construction cost index, hurricanes caused less than $400 million in damage during the 5-year period 1925–29, while the number of deaths exceeded 2,000. For the 5-year period 1960–64, hurricane damages were nearly $1.2 billion, while the same figure for 1965–69 rose to more than $2.4 billion. The dollar cost can be expected to continue climbing as greater numbers of expensive buildings are constructed in vulnerable coastal areas, and as long as inflation continues. Stormfury scientists estimate that if federal hurricane/typhoon modification research continues at at least the present level for another 10 years, and that if only one severe hurricane can be weakened so that its damage is reduced by as little as 10 percent, the federal investment will have been returned at least tenfold!

Preliminary Project Stormfury results are very encouraging. The modification experiments on Hurricane Debbie in 1969 were the first ever to "seed" a hurricane more than once per day. Debbie was seeded 5 times in an 8-hour period on both 18 August and 20 August. On 18 August, Debbie was a mature hurricane with maximum winds exceeding 100 knots. It was located about 650 nautical miles east-northeast of Puerto Rico, moving toward the west-northwest. The seeding aircraft—flying at an altitude of 33,000 feet—penetrated the hurricane eyewall to the area of maximum winds before starting to seed. Then, 208 silver iodide generators were dropped on a line extending across the outer wall of the hurricane eye and into the adjacent rainbands which help to "fuel" the hurricane. This was repeated 5 times, at 2-hour intervals. Each seeding run lasted 2 to 3 minutes over a path 14–20 nautical miles in length.

Before the first seeding on 18 August, maximum winds at 12,000 feet were 98 knots. Five hours after the fifth seeding, these winds decreased to 68 knots— a reduction of 31 percent in the maximum wind speed! On 20 August, the hurricane had a double eye, an unusual structure in Atlantic hurricanes and a complicated situation to handle with current seeding techniques. However, maximum wind speeds at 12,000 feet before the first seeding were 99 knots. After the final seeding that day, the maximum winds dropped to 84 knots—a reduction of 15 percent! The fact that the hurricane's winds were decreased considerably on both seeding days strongly suggests that at least some of the changes were caused by the Project Stormfury experiment.

In the opinion of Dr. R. Cecil Gentry, former Director of the National Hurricane Research Laboratory: "While actual modification of hurricanes may still be some years away, the rapid increase in knowledge in recent years lends support to those who believe that man will be able to exert at least some control on 'the greatest storm on earth.'"

IN MIDDLE AND HIGH LATITUDES

The traveling anticyclones (usually good weather) and cyclones (almost always bad weather) of middle and higher latitudes have already been discussed in chapter 5. The subjects of weather fronts and associated weather were treated in chapter 7. Consequently, these subjects will not be repeated here. It would be well to bear in mind, however, the average extra-tropical (or nontropical) storm tracks across the United States and into the Atlantic as shown in figure 8–23. One should also become familiar with the mean, monthly, surface-pressure charts and storm tracks for the various oceans, available from NOAA's National Weather Service at low cost.

The pressure at the center of mid- and higher-latitude cyclones of average intensity ranges from 988–1004 millibars (29.18–29.65 inches). For strong continental cyclones, the range is about 968–988 millibars (28.59–29.18 inches). For extreme cyclonic storms over the oceans, the range is about 920–968 millibars (27.17–28.59 inches). Because pressure change is closely related to wind velocity and the occurrence of bad weather, the barometer has been rightly valued for centuries as the key instrument for the early detection of storms. A pressure drop of two millibars in three hours is a warning signal of an impending storm. A drop of 5 millibars in 3 hours is strong, and a drop of 10 millibars in 3 hours is

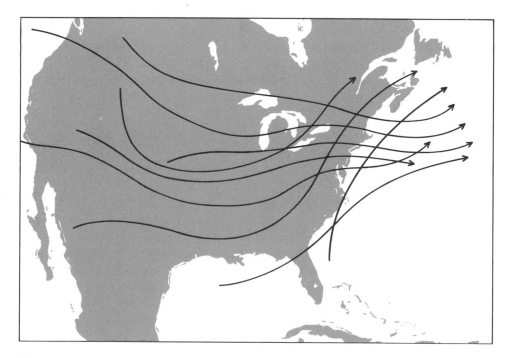

Figure 8–23. Extra-tropical storm tracks across the United States. Arrows show the paths that lows generally follow. The tracks are more widely separated in the winter than in summer.

extreme. Radical changes in weather, which are often abrupt, accompany the traveling cyclones of middle and high latitudes. The intense cyclones are accompanied by high winds and, all too frequently, weather- and/or sea-related catastrophes. At sea, one must pay particular attention not only to the actual barometric reading, but also to the value of the pressure *change* as well.

Waves in the Westerlies and Jet Stream

As a rule in temperate latitudes, the westerly winds of the middle and upper troposphere follow wavy paths. The distance from wave crest to wave crest or wave trough to wave trough can be long (3,000–5,000 miles) or short (1,000–2,000 miles) depending upon certain conditions. The atmospheric waves of short wavelength are the ones associated with active cyclone development and the generation of broad bad-weather areas. Figure 8–24 is a horizontal model of the average flow pattern between 15,000–40,000 feet. In this diagram, the distance from trough to crest is 800 miles. The horizontal mass convergence causes the

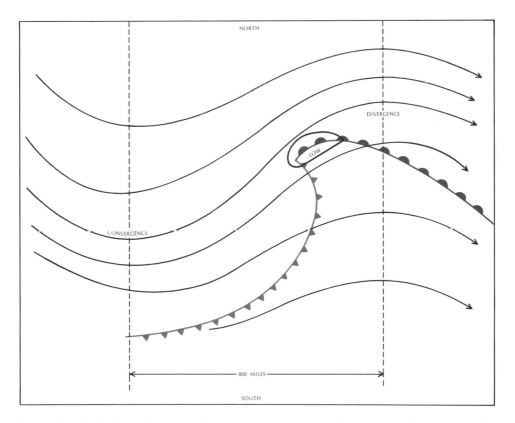

Figure 8–24. Horizontal model of the average streamline flow pattern between 15,000 and 40,000 feet over a developing surface cyclone. Surface cold and warm fronts are also indicated.

counterclockwise bending of the streamlines (lines everywhere tangent to the wind-velocity vectors) in the west. Horizontal mass divergence causes the clockwise bending of the streamlines forming the ridge to the east. The area of developing cyclone and associated surface fronts is indicated near the inflection point of the southwesterly flow. A little to the east of this point, the warm air has moved farthest northward from the tropics. (See figure 8–25.) Streamlines and isotherms (lines of constant temperature) have almost the same pattern. Often a *dome of cold air* becomes isolated from the main cold pool of air at high latitudes, as shown in figure 8–25.

One of the most interesting—and also very important—phenomena that occur in the earth's atmosphere are the *jet streams*. They have been described by various authors as "fast-moving rivers of air," "tubular ribbons of high-speed winds," "whistling winds with wanderlust," and so forth. Jet streams are usually found in the atmosphere between 30,000–40,000 feet. They average about 4 miles in height (vertically) and about 300 miles in width (horizontally). Wind speeds near the center of a jet stream—called the *core* by weathermen—are sometimes as high as 250 knots. Also, jet-stream winds are normally higher in winter than in summer because the temperature differences in the upper atmosphere, as a rule, are greater in winter than in summer.

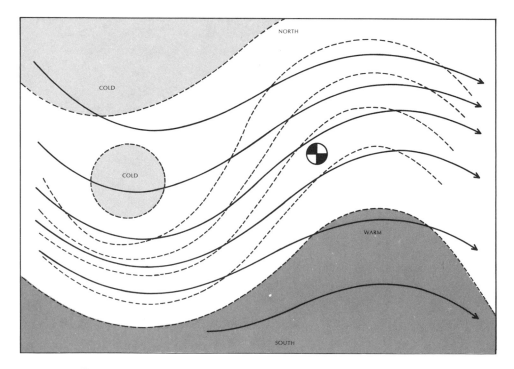

Figure 8–25. Streamlines of the upper-level flow pattern (as in figure 8–24). The dashed lines show the isotherms during the development stage of a cyclone in the middle troposphere near 20,000 feet. The checkered circle shows the position of the cyclone center.

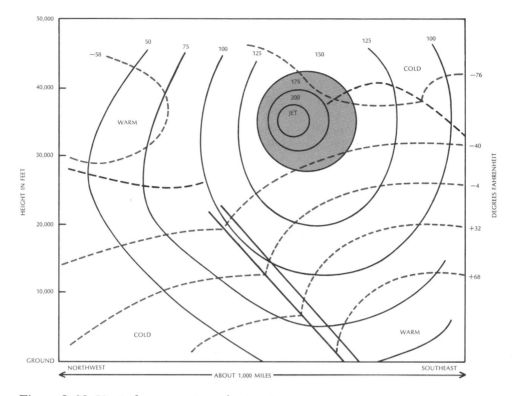

Figure 8–26. Vertical cross section of a jet stream.

Figure 8–26 is a vertical cross-section model of a jet stream. Underneath the jet stream, the temperature decreases from right to left across the stream, looking downwind. Above the jet, the reverse is true. Thus, the jet stream is centered at the height where the temperature pattern reverses with altitude. To the left of the jet stream, the tropopause is much lower than to the right. Close to the core, a distinct tropopause often cannot be found. In this tropopause "gap," the transfer of air from the stratosphere to the troposphere is thought to take place, and radioactive fallout from nuclear tests have provided the primary evidence of such a transfer. The radioactive debris often reaches the ground (from high altitudes) along a narrow belt to the right of the edge of the jet stream.

Why, you may ask, should mariners and weekend sailors be interested in an invisible weather phenomenon that occurs above 30,000 feet in the atmosphere? The reason is that the combination of short waves in the westerlies and a jet-stream core at the center of the westerly current provides the upper-air setting for the development and path of traveling surface cyclones and anticyclones. The number and intensity of jet streams and their meandering paths over and around the earth vary from day to day. And there are mid- and high-latitude westerly jets (blowing toward the east) and subtropical easterly jets (blowing toward the west). Also, rather dramatic seasonal anomalies in temperature and precipitation result from "abnormal" jet-stream meanderings and intensities.

Thunderstorms

Thunderstorms are rather spectacular and violent local phenomena produced by cumulonimbus clouds and are characterized by squalls, turbulence, strong gustiness, heavy showers, lightning and thunder, and often hailstones which can be more than four inches in diameter. The strong, gusty winds are of serious concern to surface craft, and rightly so. Visibility is invariably poor in a thunderstorm, and ceilings are low and ragged. Thunderstorms form in a number of ways, but all require warm, unstable air of high moisture content and some sort of "lifting" or "trigger" action. The air parcels must be forced upward to a point where they are warmer than the surrounding air, so that they will continue to ascend rapidly until at some point, the air parcels have cooled to the temperature of the surrounding air. The air may be lifted in several ways—by heating, by mountains or other barriers, and by fronts.

The speed at which raindrops fall depends on their size. If the drops are larger than 4 mm in diameter, their descending velocity is greater than 8 meters per second. When the drops reach this speed, they break up into smaller droplets, which then fall at a slower speed. If the rising currents in the thunderstorm are greater than 8 mps, the larger raindrops are split up into smaller droplets and are carried upward within the storm. The rising currents in thunderstorms are not steady. They actually consist of a succession of gusts and lulls. Because of this, the raindrops rise and fall, and grow and break up continuously. Each time that a drop is broken up into smaller droplets, the positive and negative electricity is separated—the air takes a negative charge and the drops assume a positive charge. By the continued splitting up of drops, enormous electrical charges are generated within the thunderstorm. Since the air rises much more rapidly than the splitting drops, the positive charge is accumulated in that part of the cloud where the rising currents are strongest, and the rest of the cloud becomes negatively charged, or neutral.

In a typical summer situation, the downdraft air arrives at the earth's surface with relatively cool temperatures, say about 67°F. The drag of the falling rain also accelerates the downdraft. When it reaches the ground, the cool air spreads out on all sides, but most rapidly in the direction toward which the thunderstorm is moving. At the approach of the cooler air, the wind dies down somewhat, and the slowly falling barometer levels off. Then a dark line of very low scud clouds approaches rapidly. The wind shifts and begins to blow with gusts of 40 to 80 knots from the direction of the thunderstorm. The temperature falls suddenly (sometimes by 20°F or more), providing some relief from the heat, and the barometer begins to rise in "jerks." All this can happen in a few minutes before the main thundercloud arrives overhead, bringing with it the heavy rain. The extreme precipitation seldom lasts more than half an hour, but in that short period of time, one and one-half to two inches of rain may fall!

It is quite easy to determine one's distance from a thunderstorm. We know that light travels at a speed of about 186,000 miles per second and that sound travels much more slowly, about one mile in a little less than five seconds (at about 1,100 feet per second). So, all one has to do is time how long it takes for the sound of thunder to reach one's position after seeing the lightning flash. Then, divide the number of seconds by 5, and one has a rough approximation of the distance to the thunderstorm, in miles. For example, if you notice a lightning flash in the afternoon at 4:25:10 p.m. and hear the thunderclap at 4:25:45 p.m., divide the 35-second time interval by the number "5" (35/5 = 7). You are about 7 miles from the thunderstorm. Figure 8–27 is a simple schematic summary of what we have just discussed.

Occasionally, most often in the spring and early summer, thunderstorm cells form a line which can stretch as far as several hundred miles, usually oriented in a north-south or northeast-southwest direction. The line may persist for 6 to 20 hours, usually traveling toward the east. There is a vast difference between an isolated thunderstorm on a hot summer afternoon and thunderstorms along a *squall line* (as discussed in chapter 7). The latter tend to be much more severe. A *mammatus* sky, described in chapter 3, usually precedes a squall line. Incessant lightning marks the arrival of a squall line, and the wind gusts may reach 100 knots. Squall lines are sometimes the forerunners of cold fronts and are to be avoided by small craft if at all possible.

Tornadoes and Waterspouts

A tornado is a circular whirl of great intensity and small horizontal extent, with winds of superhurricane force. It is the most destructive storm in the earth's atmosphere with winds up to several hundred knots revolving tightly around the center, or core. Although a tornado is only from a few feet to perhaps a mile wide, it is by far the most violent and sharply defined of nature's phenomena. Tornadoes travel along paths with lengths ranging from very short distances to 70

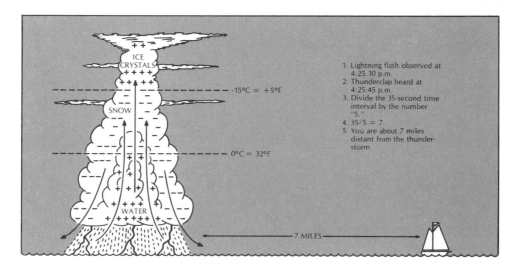

Figure 8–27. Schematic of a thunderstorm, including the distribution of electric charge. The plus signs (+) indicate a positive charge. The minus signs (−) indicate a negative charge. (After Prof. G. C. Simpson).

miles or more. The width of the path of devastation may be no more than a few city blocks. On most long paths, the funnel cloud strikes the ground and rises several times. The funnel does little damage unless it actually touches the ground. Tornadoes are usually fairly short-lived, lasting from one to several hours. They usually occur in connection with strong cold fronts which have produced squall lines. They often form in series and travel in almost parallel lines following the squall line. In the northern hemisphere, the frantic winds usually blow counter-clockwise around the vortex. Pressure near the center of a tornado has never been recorded because no meteorological installation has been devised or built which could withstand the vortex's unbelievable fury. Meteorologists are still awaiting the occasion when a tornado will pass over a house which has a barograph in the storm cellar.

Available evidence indicates that the fantastic destruction of buildings results as much from explosion outward caused by the phenomenal decrease in pressure in the center of the vortex, as from the violent winds. Consequently, the wisest course of action—contrary to one's natural inclinations—would be to open ports and hatches if a tornado were imminent. The air in the center of the vortex rises extremely rapidly, and the whirling winds are accompanied by torrential rain or hail, and thunder and lightning. The frightening sound has been compared to the roar of many locomotives. Tornadoes are most common in the spring and early summer in the U.S. when the warm and moist air masses from the Gulf of Mexico encounter the cold-air masses from the northern part of the continent. The annual mean number of tornadoes in the U.S. is 630.

Again, a *mammatus* sky often gives warning of impending tornado formation. Typically, the cloud funnel first appears near the cumulonimbus cloud base and then elongates downward. It becomes visible when condensation begins in the air that is subjected to the sudden and tremendous lowering of pressure and expansion cooling. After the funnel strikes the ground, it looks black because of the soil and debris that has been drawn into it and whirled upward.

Waterspouts are the same phenomena as tornadoes, except that they form over the sea and are less violent. When they touch the sea surface, dense spray is drawn upward into the funnel. So, one might refer to waterspouts as "wet tornadoes." It has been reported that the winds in some of the less violent water-spouts rotate clockwise. This, however, has never been observed by the writer.

Sea Breezes and Land Breezes

As we discussed previously, climates over water areas are much milder than climates over continental areas at the same latitude. Because of the influence of winds from the sea, coastal areas have higher temperatures in winter and cooler temperatures in summer than do areas farther inland. The same principle applies to larger inland lakes. In the daytime, the radiation heats the air over the land. Over water, the air temperature remains almost constant. As the air over land is warmed relative to the air over the water, the more dense (heavier) air over the water moves toward shore and proceeds inland to replace the heated, lighter air. As the heated air over the land rises, the atmospheric pressure at a height near 2,000 feet increases slightly as the air expands, and the air at this height is accel-erated toward the water, as shown in figure 8–28. As the surface pressure over

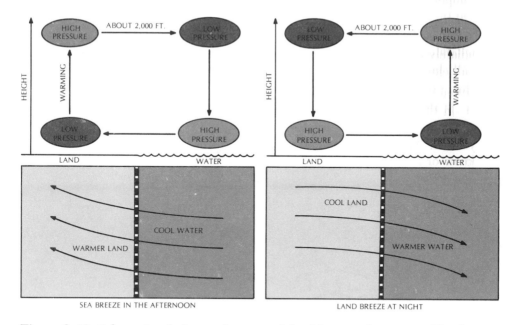

Figure 8–28. Schematic of the sea-breeze and land-breeze phenomena. The bottom diagrams are horizontal views. The upper diagrams are vertical cross sections.

the land drops, the air near the ground is accelerated from the water toward the shore. In this way, sea breezes and lake breezes are generated (from water toward land). Sea breezes sometimes reach speeds in excess of 15 knots and can be much stronger when they are in the same direction and reinforcing the "normal" wind caused by the existing pressure distribution. When the two winds are in opposite direction, the resultant wind is the vector difference between the two. On occasion, sea breezes have penetrated inland as far as 50 miles. At nighttime, when the air over the land has cooled more than the air over the water, the reverse process takes place. The wind then blows from the cooler land toward the warmer water, and this is known as a land breeze (from land toward water). Sea breezes and land breezes are important to know about when sailboating in coastal areas or on large lakes.

HURRICANE-FORCE WINDS OUTSIDE THE TROPICS

Hurricanes and *typhoons* are usually understood to be severe storms from the tropics. But for all practical purposes—and especially from a seagoing man's point of view—this definition is too limited. The threshold of hurricane- and typhoon-force winds is 64 knots (74 mph). Winds of such strength occur many times over the oceans each winter season in cyclones well outside the tropics.

As mentioned earlier, one of the main differences between tropical and extra-tropical cyclones is that the tropical cyclones have a warm core, or center, whereas the extra-tropical cyclones have a cold core. Another difference is that the extra-tropical cyclones usually have a much larger diameter. But perhaps the most important difference is the distribution of wind speeds (isotachs) around the two types of vortices.

In a tropical (warm core) cyclone, the hurricane-force winds exist in the area immediately around the center, decreasing in strength as one goes outward from the wall cloud of the eye. But in an extra-tropical cyclone (cold core), the hurricane-force winds almost always exist in various shapes or patterns away from the center—in the storm's periphery. Figure 8–29 is a schematic diagram showing a comparison of the distribution of hurricane-force winds around the two types of cyclones. The shaded area represents the regions of hurricane-force winds (64 knots or higher).

The areas off Cape Hatteras, North Carolina, the Icelandic area of the North Atlantic, and the Aleutian area of the North Pacific, are favorite regions for the

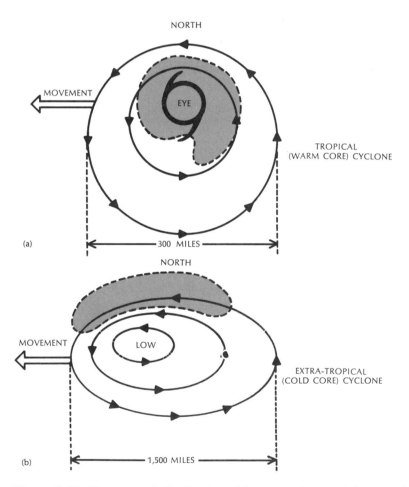

Figure 8–29. The general distribution of hurricane-force winds around tropical (warm core) and extra-tropical (cold core) cyclones. The shaded areas represent the regions of hurricane-force winds (64 knots or higher).

development of hurricane-force winds which generate tremendous ocean waves during the winter months. These winds sometimes blow without appreciable change in direction or speed for a distance of 1,500 miles or more, persisting for 48 to 72 hours. Over such a long distance, mountainous waves are created by the prolonged winds. Marinas, other coastal installations, homes, and other real estate are periodically exposed to the tremendous and highly dangerous force of these waves, swells, and exceptionally high tides. Who can ever forget the devastation along the entire east coast of the North American continent, extending from Canada to Florida, wrought by the storm of 6–9 March 1962?

9
forewarning and formation of fog

"The fog comes on little cat feet."
                    —Carl Sandburg

One of the most interesting, and in some respects exceedingly important, weather phenomena is *fog*—a great swarm-like assemblage in the surface air of hundreds of thousands of water droplets so minute that it would take seven billion of them to fill a teaspoon. Expressed in the most simple terms, fog is merely a cloud whose base rests on—or very close to—the earth's surface, reducing the visibility to five-eights of a mile or less. Although the substance of fog and cloud is exactly the same, the processes of cloud and fog formation are quite different. Clouds form chiefly because air rises, expands, and cools. Fog results from the cooling of air which remains at the earth's surface.

Fog is not only a nuisance and a menace to navigation, it has its lethal aspects, as well. During the London fog of 27 November through 1 December 1948, thousands of cars had to be abandoned by their drivers, trains and flights were canceled, buses were led by their conductors on foot with flashlights and lanterns, and marine shipping in the Thames was at a standstill. Football matches also had to be canceled in the zero visibility. It was estimated that this fog caused between 700 to 800 additional deaths in London. In the "killer-fog" of London from 5–9 December 1952, an estimated 4,000 people died of bronchitis and pneumonia complications as a direct result of what is now known as smog.

THE DEW POINT

Before we get into a discussion of the different types of fogs and their formation, we must say a word about *dew point* (or *dew-point temperature*). Specifically, the dew point is the temperature to which a given parcel of air must be cooled—at constant pressure and constant water vapor content—in order for saturation to occur. When this temperature is below 32°F, it is sometimes called the *frost point*. Most weathermen prefer to define the dew point as the temperature at which the saturation vapor pressure of a given parcel of air is equal to the actual vapor pressure of the contained water vapor. But why complicate things? Bear in mind the former definition and remember the following: (1) when there is a large difference between temperature and dew point and the difference does not decrease, there will be no fog; and (2) when the difference between the temperature and dew point decreases and the two values approach each other, or the difference becomes zero, look out for fog.

Let us take this point further. The curve in figure 9–1 represents the amount of water vapor required to saturate the air at sea level for any given temperature. Note that at higher temperatures, the air—or more accurately, the space occupied by the air—can hold more water in the vapor state. The air along each point of the curve is saturated with water vapor and is at its dew point. Any further cooling will yield water as a result of condensation. Hence, fog or low-ceiling clouds will form—depending on the wind velocity. As the cooling continues, the dew point will fall because more and more vapor has condensed.

Raising the dew point is usually accomplished by the evaporation of water from falling precipitation or by the passage of a body of air over a wet surface. A high dew point is characteristic of maritime air masses. Continental air masses that travel slowly over such water surfaces as the Great Lakes show a considerable change in dew point. It can be seen from figure 9–1 that at higher tempera-

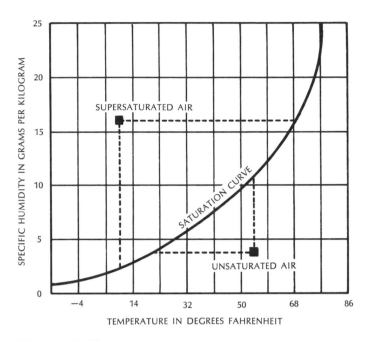

Figure 9–1. The saturation curve.

tures a very small increase in temperature means that much more water vapor can be accommodated. A much larger change in temperature would have to be made at colder temperatures to accomplish the same increase in water content. This means that more heating by the sun is required to dispel a fog (or low-ceiling cloud) at 40°F than one at 60°F, even though the liquid content of the fog (or low-ceiling cloud) may be the same in both cases. It takes longer for a fog to dissipate in winter than in summer, partly for the reason just given and partly because there is less insolation in winter than in summer.

While stratus and cumulus cloud formations are preceded by cooling under the conservation of heat, the formation of fog involves the actual withdrawal of heat from the air, mainly through radiation cooling and the movement of air over a cold surface.

Radiation Fog

Radiation fog is a nighttime, over-land phenomenon, which occurs when the sky is clear, the wind is light, and the relative humidity is high near sunset. When the relative humidity is high, just a little cooling will lower the temperature to the dew point. Strong winds prevent nocturnal cooling. In fact, the turbulence generated by winds of only 15 knots will hinder fog formation completely. A light wind of three knots, however, is favorable for the formation of fog. The very small amount of turbulence it generates mixes the air particles which are cooled at the ground upward, thus ensuring a solid fog layer up to 40 to 100 feet. In absolutely calm air, radiation fog will be patchy and may be only waistdeep.

Because cold air drains downhill, radiation fog is thickest in valley bottoms—with the surrounding hillsides rising above the fog. Although radiation fog does not occur over water surfaces (because the heating qualities of water are such as to hold the day-to-night temperature range of water surfaces to 3°F, or less), it obscures the shore-based beacons and landmarks, and complicates navigation. Radiation fog normally disappears about one to three hours after sunrise. If it is mixed with smoke, which is especially common in winter, it forms a greasy *smog* (*sm*oke and f*og*) which may not be dissipated until late morning.

Autumn and winter are the most favorable seasons for radiation fog. During the autumn, moisture content is still high in the air, and during the winter, the nights are longest. The center of a high-pressure area is a favorite spot for radiation fog because the winds are light and the skies are usually clear (chapter 5).

Advection Fog

Advection fog may form—day *or* night—when warmer air blows over a colder land or water surface. We say the air is *advected*. This case is entirely different from radiation fog. In the case of advection fog, the warmer air gives off heat to the colder underlying surface, and this cools the air to the dew point. As the cool surface chills the warm air flowing over it, the water vapor of the air tends to condense on particles of salt, dust, or smoke, even before the relative humidity reaches 100 percent. The fog will be light, moderate, or dense, depending on the amount of water vapor that condenses.

The air in which fog forms does not tend to rise because it is being cooled, thus becoming more dense, heavier, and more stable. However, if the wind speeds are 15 to 20 knots or more over land surfaces, the air is mixed through a relatively deep layer and individual particles do not remain long enough at the earth's surface to be cooled to, or near, the dew point. Low stratus or stratocumulus clouds, rather than fog, form under these conditions. Over the sea, however, where there is less frictional turbulence than over land, advection fog will form even when the wind speeds are 30 knots. It is usually extensive and persistent.

The temperature of the earth's surface (in the northern hemisphere) normally decreases as one goes northward. Thus, advection fogs form mainly when the air currents have a direction component from the south, especially when the currents have a high moisture content.

Water areas are relatively cold for the following reasons:
1. The presence of a cold water current, such as the Labrador Current across the North Atlantic (see figure 9–2).
2. A cold upwelling of water, such as is found off the northern California coast (figure 9–3).
3. Contrasting water temperatures, such as are found along the Aleutian Islands chain where the Bering Sea is colder than the North Pacific (see figure 9–4).
4. The drainage of cold river water into a large, relatively warm saltwater area. Since fresh water is usually less dense than salt water, the less-dense, cold fresh water remains on top. This sometimes establishes a cold water area over scattered parts of the northern section of the Gulf of Mexico during winter months. When water temperatures are colder than the dew point of the warm air, fog will form (see figure 9–5).

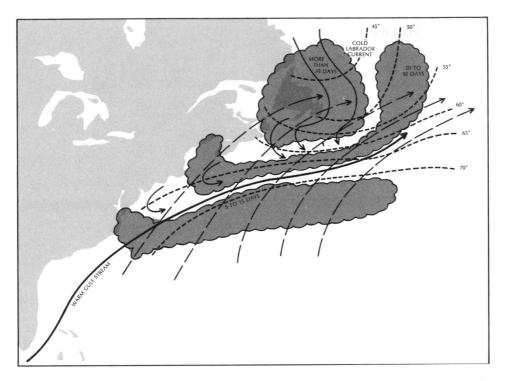

Figure 9–2. Fog over the western North Atlantic during the summer months. The shaded areas are fog; numbers in the middle refer to the average number of days with fog during the period June-August. Dashed lines are sea-surface temperature (isotherms) in degrees Fahrenheit. Dashed arrows are streamlines of the prevailing surface air currents. Solid arrows represent the cold Labrador Current and the warm Gulf Stream.

Tropical-Air Fog

Tropical-air fog is really a form of advection fog which does not depend upon the flowing of warm air over a cold current, but rather on the *gradual* cooling of the air as it travels from lower to higher latitudes. It occurs over both water and land and is probably the most common type of fog over the open sea. In the United States, it forms some of the most widespread fogs observed anywhere. Tropical-air fog is more common over water than over land because of the smaller frictional effect over water. See, again, figure 9–2.

Frontal Fog

As mentioned briefly in chapter 7, fog sometimes forms ahead of warm fronts and occluded fronts and behind cold fronts. This happens when the rain falls from the warm air above the frontal surfaces into the colder air beneath. Evaporation from the warm raindrops causes the dew point of the cold air to rise until condensation on the nuclei present takes place.

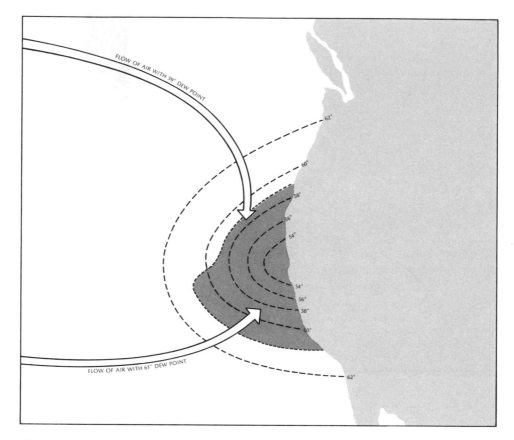

Figure 9–3. Northern California coastal fog resulting from upwelling of cold water. Shaded area is fog. Dashed lines are sea-surface temperature lines (isotherms) in degrees Fahrenheit.

Upslope Fog (Expansion Fog)

When air travels upslope for a long time over rising terrain, the decrease in temperature which it undergoes because of adiabatic expansion may lead to fog. Across the great plains of the United States, the standard atmosphere pressure decreases by about 130 millibars from near the Mississippi to the eastern edge of the Rockies. Surface air traveling this route from east to west undergoes a temperature decrease of about 23°F. This westward motion occurs mainly in late winter and spring. At that time, the difference betweeen temperature and dew point in the lower plains is sufficiently below 23°F, so that the air traveling over the slowly rising terrain will form upslope (or expansion) fog about 100 to 200 miles to the east of the mountains.

Steam Fog and Arctic Sea Smoke

During the early morning hours in summer, long columns of steam, or mist, often rise over inland bodies of water, lakes, and even river valleys. At that time of year, water temperatures and vapor pressure over water are at their highest. Cold air

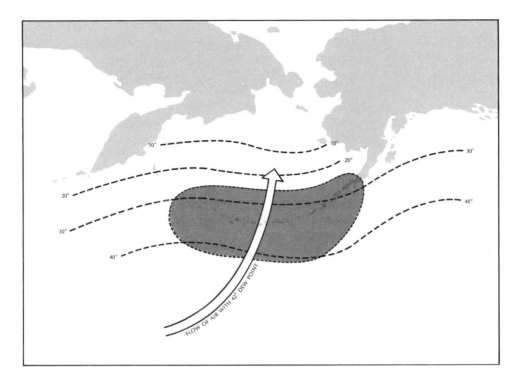

Figure 9-4. The famous Aleutian fog that persists even when the winds reach 35 knots. The shaded area is fog. Dashed lines are sea-surface temperature lines (isotherms) in degrees Fahrenheit.

draining down the slopes may be 18°F or more, colder than the water. The water evaporating from the surface may supersaturate this air almost immediately. The evaporated water recondenses and rises with the air that is heated from below. Steam fog is usually very shallow and looks like tufts of whirling smoke coming out of the water.

When very cold air in winter moves from a land mass out over water which is perhaps 45°F warmer than the air, or when arctic air travels from the ice shelves onto the open ocean, the steaming is so intense that it consolidates into a *fog* and is called *arctic sea smoke.*

Inversion Fog

Inversion fog is typical of the subtropical west coasts of continents—California, for example. The upwelling of cold water is common along west coasts because of the wind action, and the air flowing over this cold water acquires a low temperature and high relative humidity. Above the cool, moist layer of air is a temperature inversion (increase of temperature with height) which acts as a "lid" and prevents the moist air from rising. See, again, figure 9-3. Fog forms, similar to the advection process, and is very frequent offshore. At night, as the land cools, the fog works its way inland.

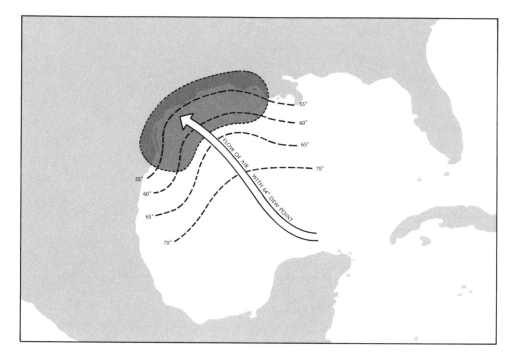

Figure 9–5. Fog caused by the drainage of cold river water into the Gulf of Mexico during the winter months. Shaded area is fog. Dashed lines are sea-surface temperature lines (isotherms) in degrees Fahrenheit.

Ice Fog

Ice fogs occur at temperatures of −22°F or colder, mostly in inhabited areas. Aircraft running up their engines in temperatures of −40°F or lower will quickly cause an airport to be enveloped with ice fog. Even animals, after a bit of exercise, find themselves surrounded with ice fog caused by the evaporation of body liquids which condense and freeze in the bitterly cold air.

Fog Over the Sea

Fog is rare at the equator and in the trade-wind belts, except along the coasts of California, Chile, and northwest and southwest Africa. But it is a common phenomenon in the middle and high latitudes, especially in spring and early summer. The Gulf Stream and the Labrador Current converge in the Newfoundland Banks region. When warm air from the Gulf Stream overruns the cold water of the Labrador Current, dense fog banks result. Similarly, in the northwest Pacific, fog is common off the coast of Asia where the warm Japanese Current overruns the cold Kamchatka Current. During the spring and summer months, warm and moist air currents which flow from land to sea produce fog over coastal water areas. A shift in wind direction tends to drift the fog back over the adjacent land areas. During the fall and early winter months, the air currents blowing from sea to land tend to produce fog over coastal sections. Such fog may drift to sea with a reversal of the wind direction.

In the North Pacific Ocean and Bering Sea, broad areas of dense fog are common, and at times the fog extends to heights of 5,000 feet or more. It is caused by air moving northward from the high-pressure areas of the Pacific Ocean centered near latitudes 35 to 40 degrees north. The air is cooled to its dew point in passing over the colder waters to the north. This cooling of the air also increases its density, so that there is little tendency of the air to rise. See, again, figure 9–4. Fog persists in the Aleutian area even when winds are 35 knots. During the fall and spring months, fog is at a minimum in the Aleutian area. In winter, arctic sea smoke is quite common, and at times it may build upward to several thousand feet thick.

Forewarning of Fog

Perhaps the most important single factor to remember in connection with fog is the difference between air temperature and dew point—called the dew point *spread*. Whenever the formation of fog is anticipated or predicted, it is well to keep a plot of this difference (the spread), and watch the trend closely. As the spread approaches zero, fog is imminent. Table 9–1 contains a summary of the fog-producing and fog-dissipating processes in the atmosphere. Figures 9–6 through 9–11 are classic examples of some of the different types of fog.

Table 9–1. Summary of the Fog-Producing and Fog-Dissipating Processes in the Atmosphere.

Fog-Producing Processes	*Fog-Dissipating Processes*

A. Evaporation from:
  1. Rain that is warmer than the air (frontal fog and rain-area fog).
  2. Water surface that is warmer than the air (steam fog).
	A. Sublimation or condensation on:
	  1. Snow with an air temperature below 32°F (except ice-crystal fog).
	  2. Snow with an air temperature above 32°F (melting snow).

B. Cooling from:
  1. Adiabatic upslope motion (upslope fog).
  2. Radiation from the underlying surface (radiation fog).
  3. Advection of warmer air over a colder surface (advection fog).
	B. Heating from:
	  1. Adiabatic downslope motion.
	  2. Radiation absorbed by the fog or by the underlying surface.
	  3. Advection of colder air over a warmer surface.

C. Mixing:
  1. Horizontal mixing (unimportant by itself and strongly counteracted by vertical mixing).
	C. Mixing:
	  1. Vertical mixing (important in the dissipation of fogs and the production of stratus clouds).

Figure 9–6. Ground fog. Courtesy of Hunting AeroSurveys, Ltd.

Figure 9–7. Radiation fog in the evening. The photograph was taken after the lights were turned on, showing the patchiness and variation of the fog with height. Courtesy of R. K. Pilsbury.

Figure 9–8. Sea fog. The fog is less dense in the warmer waters of the bay on the left, but quite thick on the right, especially where the air is forced to rise over the hills. Courtesy of R. K. Pilsbury.

Figure 9–9. A broad belt of fog forming as relatively warm, moist air is forced to rise over the hills. Note that conditions are still quite clear at sea level, but not for long. Courtesy of R. K. Pilsbury.

Figure 9–10. Fog forming in a valley below the level of the observer. Courtesy of R. K. Pilsbury.

Figure 9–11. Fog rising and clearing from a valley. Note the individual puffs of fog in the sun's rays to the left. Courtesy of R. K. Pilsbury.

10
ice accretion - hazard of significance to seafarers

"To a child, all weather is cold."
— J. Heywood, 1546

The accretion of ice on ships and small craft in cold waters has not, until very recently, been given a great deal of thought by individuals other than those directly affected by it. Yet, it constitutes an especially serious hazard to the smaller ships and small craft. But even for the larger ships, including warships, the accumulation of ice on deck, on superstructure, and on equipment, greatly impairs their overall efficiency—and sometimes their safety.

The added weight of ice reduces freeboard, and therefore reduces the range of stability of the vessel. Even more dangerous, however, is the ice formed high on masts, rigging, and superstructure. This ice has a large heeling lever, and has a most serious effect on stability. The vessel may become top-heavy and capsize. Compounded with the danger from the loss of stability is the effect of the windage, or "sail area," of the ice on masts and rigging which may make the ship or small craft difficult to handle in heavy weather. For example, it may become impossible to stay head to the wind. In addition, the accumulation of ice on aerials may render the radio and radar inoperative. It is now well known that many trawlers have been lost at sea because of ice accretion.

THE NATURE OF ICING AT SEA

Three of the basic mechanisms of ice accretion are: (1) freezing rain, (2) arctic frost smoke, and (3) freezing spray. Each of these will be discussed very briefly in the following sections.

Freezing Rain

Freezing rain will cover a ship or small craft with fresh-water glaze ice, but the accumulated weights of ice are unlikely to be sufficient to endanger the vessel directly. Rates of ice accretion can be expected to be commensurate with rates of accumulation of rainfall, and these are not great, especially in the colder climates.

Arctic Frost Smoke

As mentioned in chapter 9, arctic sea smoke occurs when the air is at least 16°F colder than the sea. If the air temperature is below 32°F, then the arctic sea smoke is called *arctic frost smoke*. This frost smoke is often confined to a layer only a few feet thick, and trawlermen in northern waters refer to it as *white frost* when the top of the layer is below the observer's eye level. It is referred to as *black frost* when it extends above the observer.

The small water droplets in the frost smoke are supercooled. On contact with the vessel, part of the droplet freezes immediately, while the remainder stays liquid for a short time before it, too, freezes in the cold air. The result of the instantaneous freezing of the supercooled droplets is an accretion of opaque, white rime ice with imprisoned air. This rime ice is easier to remove than the clear ice, or glaze, which forms in other circumstances, because rime ice is porous.

Occurrences of icing on the British fisheries research vessel *Ernest Holt* have been reported in great detail. In one case, about 100 miles east of Bear Island, the ship was in air temperatures colder than 14°F, with dense frost smoke. In about 12 hours, 4 inches of rime ice had accumulated on deck, with a 12-inch thickness of ice on the ship's side at the level of the rail. It was calculated that the rate of accretion was as much as two and a half tons per hour in this case!

Freezing Spray

Freezing spray is the most dangerous form of icing. It occurs when the air temperature is below the freezing temperature of the sea water, about 27°F. The spray freezes on the exposed surfaces of the vessel to produce clear ice or glaze. At lower air temperatures, the ice may be opaque and this may be due to the spray being supercooled so that it partially freezes on impact. At extremely low temperatures, such as 1°F and below, as might be encountered in anchorages or close inshore, wind-induced spray may be frozen before it strikes the vessel, so that it does not adhere to the vessel, but may form drifts on deck. This would not necessarily apply to any spray generated by the movement of the vessel through the water, which would strike the vessel while still liquid, and adhere. Sea ice formed from green water coming inboard will seriously add to the accumulation of ice if it is trapped on deck by ice-choked rails or freezing ports.

With air temperatures below 28°F, freezing spray is observed in winds of 18 knots or higher. The lower the air temperature and the stronger the wind, the more rapid is the accumulation of ice. A low sea temperature also increases the rate of accumulation of ice. To the spray blown from the wave caps is added the spray generated by the vessel herself, so that the total rate of ice accretion will also depend on the design and loading of the vessel, on her heading and speed relative to the waves, and also on the relative wind (which determines which part of the vessel is most exposed). It should be understood that an accumulation of ice will, itself, increase the rate of accumulation, since the ice already formed increases the effective cross section of rigging, mast, rails, and so forth, exposed to the spray. Thus, there are just too many variables for simple, precise rules to be formulated. However, figures 10–1 through 10–4 will prove a very useful guide in your forecasting the relative severity of icing conditions as determined by weather factors. These figures are based on the outstanding research in this field by Dr. H. O. Mertins.

It is interesting to note that at the investigation conducted into the loss of the trawlers *Lorella* (559 tons) and *Rodbrigo* (810 tons), the ice accretion from freezing spray on these ships amounted to 50 tons or more in 24 hours, and this was sufficient to cause them to founder in the rough seas. Weather conditions at the time were estimated to be as follows: (1) air temperature 25°F, (2) seawater temperature 34°F, and (3) wind speeds near 50 knots. As a matter of interest, plot these conditions on figures 10–3 and 10–4.

No icing has been observed where the sea-surface temperatures were above 41°F. There have been cases of icing with sea-surface temperatures between 37° to 41°F, but the accretion of ice did not constitute a real threat.

Japanese experiments in the western North Pacific have confirmed the accretion of ice at wind speeds of 18 knots and above, with air temperatures colder than 28°F. Also, some ice accretion even occurred in wind speeds of only 13 knots. The relationship between the rate of accretion of ice and the various weather factors was not determined, but a particular accumulation measured was two tons per hour (on a 400-ton ship) in a wind of 18 knots with the air temperature 20°F and the sea temperature 32°F. In the course of these experiments, accumulations of ice up to 26 tons were frequently observed, and in one case, an accumulation of 60 tons of ice on a 500-ton ship in a period of only 24 hours was observed!

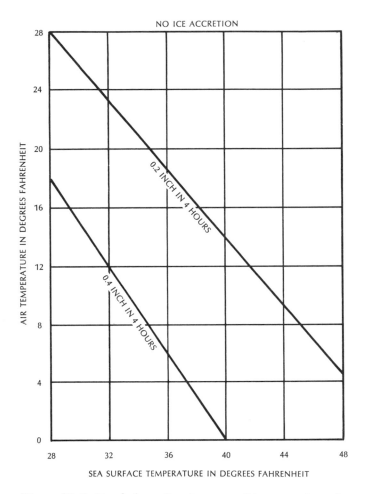

Figure 10–1. Graph for estimating rate of ice accretion when wind speeds are between 22 and 34 knots.

METEOROLOGICAL ASPECTS AND PREDICTIONS

Despite the many uncertainties about the quantitative aspects of ice accretion on vessels, it is evident that the worst conditions are met in the combination of very low air temperatures and strong winds. Although other circumstances also give rise to these conditions, the weather features characteristically producing them are found to the rear of a low-pressure system and on its poleward side. Since cold air will be warmed by passing over warmer water, the coldest air will be encountered when the wind is blowing directly off a cold land area (or ice).

Because of the uncertainties concerning the rate of accretion of ice, and also because it depends, in part, on such factors as vessel design, and course and speed relative to both waves and wind, it must be realized that the wording of any predictions relating to the rate of ice accretion will be quite difficult for the weatherman, and any scale of intensity used will necessarily be a coarse one.

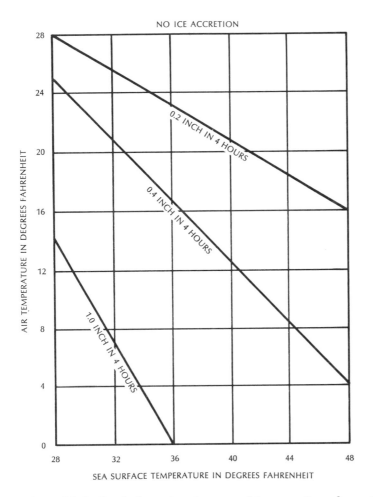

Figure 10–2. Graph for estimating rate of ice accretion when wind speeds are between 34 and 40 knots.

AVOIDING THE HEAVIEST ICING

Complete protection from shipboard icing involves avoiding areas of strong winds where the air temperatures are below 28°F. When the avoidance of a strong-wind area is not practicable, then the severity of icing can be reduced by heading for less cold air and warmer water. As can be seen from figures 10–1 through 10–4, the sea-surface temperature has a direct influence on the rate of ice accretion. In addition, warmer sea temperatures frequently mean warmer air temperatures. Since the air temperature also has a strong influence on the rate of ice accretion, the benefit of warmer water may be two-fold.

Vessels that seek shelter in the lee of land may still experience low air temperatures, but there should be some reduction in wind speeds and less spray blown from the wave caps. Ship-generated spray in the calmer water will be greatly reduced. Furthermore, attempts to remove the ice will not be thwarted

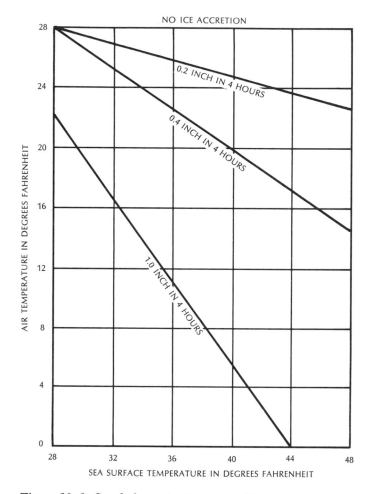

Figure 10–3. Graph for estimating rate of ice accretion when wind speeds are between 41 and 55 knots.

by the vessel's motion and seas sweeping the deck, as would be the case in the open sea.

In very high latitudes, however, do not seek shelter in the lee of the ice edge. The ice provides negligible shelter from the wind, and here the coldest air and sea temperatures are found, and provide the most severe conditions for icing. If the wind backs or veers parallel to the ice edge, the air temperature remains very low and heavy seas are soon generated.

EQUIVALENT CHILL TEMPERATURE

The risk of freezing one's flesh is not to be taken lightly, and temperature is only one consideration. At the same temperature, the danger of freezing exposed flesh (face, ears, nose, hands, etc.) increases as the wind speed increases, up to a

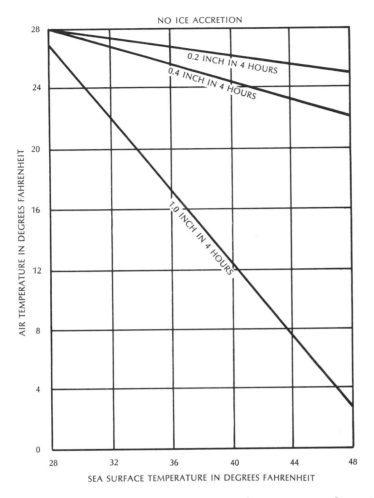

NO ICE ACCRETION

0.2 INCH IN 4 HOURS

0.4 INCH IN 4 HOURS

1.0 INCH IN 4 HOURS

AIR TEMPERATURE IN DEGREES FAHRENHEIT

SEA SURFACE TEMPERATURE IN DEGREES FAHRENHEIT

Figure 10–4. Graph for estimating rate of ice accretion when wind speeds are 56 knots or greater.

maximum of about 36 knots. Above that speed, winds have little additional effect. Under certain conditions of temperature and wind, flesh may freeze within 30 seconds! Considerable research in this field has been accomplished by the U. S. Air Force in Alaska.

The cooling power of the wind is expressed as the *equivalent chill temperature*. Knowing (or estimating) what the temperature and wind speed are, one can use Table 10–1 to determine the equivalent chill temperature—that is, the temperature that would cause the same rate of cooling under calm conditions. Referring to Table 10–1, assume a temperature of 10°F and a wind speed of 30 knots. What is the equivalent chill temperature? In other words, what effect does that combination of wind and temperature have upon exposed flesh? Going horizontally across the top of Table 10–1 stop at the block labeled 10°F. Then go down vertically until you are opposite the 29–32 knots in the wind speed column

Table 10-1. The cooling power of the wind expressed as "Equivalent Chill Temperature," and the danger of freezing exposed flesh. Courtesy of USAF Air Weather Service (MAC).

Cooling Power of Wind Expressed as "Equivalent Chill Temperature"

Wind Speed Knots	MPH	Temperature (°F)																				
		40	35	30	25	20	15	10	5	0	-5	-10	-15	-20	-25	-30	-35	-40	-45	-50	-55	-60
Calm	Calm	40	35	30	25	20	15	10	5	0	-5	-10	-15	-20	-25	-30	-35	-40	-45	-50	-55	-60
3-6	5	35	30	25	20	15	10	5	0	-5	-10	-15	-20	-25	-30	-35	-40	-45	-50	-55	-60	-70
7-10	10	30	20	15	10	5	0	-10	-15	-20	-25	-35	-40	-45	-50	-60	-65	-70	-75	-80	-90	-95
11-15	15	25	15	10	0	-5	-10	-20	-25	-30	-40	-45	-50	-60	-65	-70	-80	-85	-90	-100	-105	-110
16-19	20	20	10	5	0	-10	-15	-25	-30	-35	-45	-50	-60	-65	-75	-80	-85	-95	-100	-110	-115	-120
20-23	25	15	10	0	-5	-15	-20	-30	-35	-45	-50	-60	-65	-75	-80	-90	-95	-105	-110	-120	-125	-135
24-28	30	10	5	0	-10	-20	-25	-30	-40	-50	-55	-65	-70	-80	-85	-95	-100	-110	-115	-125	-130	-140
29-32	35	10	5	-5	-10	-20	-30	-35	-40	-50	-60	-65	-75	-80	-90	-100	-105	-115	-120	-130	-135	-145
33-36	40	10	0	-5	-15	-20	-30	-35	-45	-55	-60	-70	-75	-85	-95	-100	-110	-115	-125	-130	-140	-150

Equivalent Chill Temperature

Winds Above 40 Have Little Additional Effect

Little Danger

Increasing Danger
(Flesh may freeze within 1 min.)

Great Danger
(Flesh may freeze within 30 seconds)

Danger of Freezing Exposed Flesh for Properly Clothed Persons

at the left of the figure. Note that the combination of 10°F temperature and a 30-knot wind have the same effect upon exposed flesh as a temperature of −30°F under calm conditions. Under these circumstances, flesh may freeze within one minute. Using another example, the combination of a temperature of −20°F and a wind of 22 knots yields an equivalent chill temperature of −75°F, and flesh may freeze within 30 seconds. For any combination of temperature and wind, enter Table 10–1 at the closest 5° interval along the top, and with the appropriate wind speed along the left side. The intersection gives the approximate equivalent wind chill temperature.

It is important to remember that even though the wind may be calm, the freezing danger is great if a person is exposed on the deck of a ship or craft, in a moving vehicle, under helicopter rotors, in a propellor blast, and so forth. It is the speed of *relative air movement* that is important, and the cooling effect is the same whether one is moving through the air, or the air is blowing past a person.

The danger of freezing exposed flesh is less if an individual is active. An individual produces about 100 Watts (341 British Thermal Units—BTUs) of heat when standing still. But in vigorous activity and exercise, an individual produces up to 1,000 Watts (3413 BTUs). An equivalent chill temperature chart should always be taken on long, cold cruises.

11
wind, waves, and swell

It is pleasant, when the sea is high and the winds
are dashing the waves about, to watch from
shore the struggles of another.
                    —Lucretius, 99–55 B.C.

Mankind has been greatly interested in the oceans since before the time of Aristotle, who wrote a treatise on marine biology in the fourth century B.C. But man's interest in the sea is not surprising when one considers that the volume of the earth's oceans is something like 328 million cubic miles, and that the volume of all land above sea level is only 1/18th of the volume of the oceans. The early studies of the ocean were concerned with problems of commerce. Information about tides, currents, and distances between ports was required.

When he was Postmaster General, Benjamin Franklin prepared temperature tables by means of which navigators could determine whether or not they were in the Gulf Stream. This resulted in faster mail service to Europe. However, the beginning of modern oceanography is usually considered to be 30 December 1872 (over 100 years ago), when HMS *Challenger* made her first oceanographic station on a three-year round-the-world cruise. This was the first purely deep-sea oceanographic expedition ever attempted. Analysis of the sea water samples collected on this expedition proved for the first time that the various constituents of salts in sea water are virtually in the same proportion everywhere. But even before the *Challenger* expedition, Lieutenant Matthew Fontaine Maury of the U.S. Navy (often called the father of American oceanography) was analyzing the log books of sailing vessels to determine the best oceanic routes. He did much to stimulate international cooperation in oceanography and marine meteorology.

The question is frequently asked, "What is the difference between hydrography and oceanography?" Perhaps the best way to answer this is to compare the ocean to a bucket of water. Then, hydrography would be a study of the bucket, and oceanography would be a study of the water. Hydrographers are concerned primarily with the problems of navigation. They chart coastlines and bottom topography. A hydrographic survey usually includes measurement of magnetic declination and dip, tides, currents, and weather elements. Oceanography is concerned with the application of all physical and natural sciences to the sea. It includes the disciplines of physics, chemistry, geography, geology, biology, and *meteorology*. The interrelationship of specialties is one of the main characteristics of oceanography, and many times the words *hydrography* and *oceanography* are used interchangeably.

Although the subject of oceanography has many fascinating aspects, we will necessarily limit our discussion to wind-generated waves and swell—perhaps the greatest single oceanographic hazard to navigation in coastal waters and on the open sea.

WAVES, SEA, SWELLS, AND BREAKERS

Professional mariners live in intimate contact with the waves of the sea and are able to realize better than most people the extent to which wave size and energy (as demonstrated by the destructive power of waves) are related to the speed of the wind. They are also accustomed by their training and experience to make frequent estimates of the surface wind speed by noting the appearance of the sea surface. For the amateur mariner or weekend sailor, this is not so easy. Figures 11–1 through 11–7 show the appearance of the surface of the sea at graduated wind speeds ranging from 8 to 90 knots and constitute an excellent wind-speed/

Figure 11–1. Wind at 8 knots. Sea has occasional whitecaps. Official U.S. Navy Photograph.

Figure 11–2. Wind at 12 knots. Occasional small whitecaps are elongated, perpendicular to the wind direction, and carry downwind a short distance. Official U.S. Navy Photograph.

Figure 11–3. Wind at 25 knots. Larger elongated waves begin to form. Numerous patches of white foam remain from breaking waves. Wind streaks become more noticeable. Official U.S. Navy Photograph.

Figure 11–4. Wind at 35 knots. Light spray begins to blow off the breaking waves and is carried downwind. Official U.S. Navy Photograph.

Figure 11–5. Wind at 45 knots. Wind streaks become continuous. Official U.S. Navy Photograph.

Figure 11–6. Wind at 60 knots. Entire sea takes on a whitish-green cast. Official U.S. Navy Photograph.

Figure 11–7. Wind at 90 knots. Sea becomes whiter and whiter from blowing spray and foam. Official U.S. Navy Photograph.

sea-state catalog with which the reader should become thoroughly familiar—especially if there is no anemometer (an instrument which indicates wind speed and direction) on board. Figures 11–1 through 11–7 were taken from Navy aircraft. For a comparison with photographs taken from a ship, see Figures 11–8 through 11–20, which represent the Beaufort Scale.

There is a large distinction between waves, sea, swells, and breakers. First of all, contrary to popular belief, waves are *not* masses of horizontally traveling water particles. The wave *shape* moves horizontally, but the individual water particles—except in cases where large friction is involved—describe orbital circular motions, but remain essentially in place with little forward movement. If we throw a cork over the side, we notice that although the waves move, the surface of the water does not travel with the waves. Instead, the cork—and every other point on the water surface when waves are present—describes a circle in a vertical plane, moving upward as the wave crest approaches, forward as the crest passes, downward as the crest recedes, and backward as the trough passes. The diameter of this circle is equal to the height of the waves, and the time in which the cork moves once around it is equal to the *period* of the waves. Similarly, the movement of every point on the surface of a water wave takes place in a vertical circle. Thus, the profile of the wave surface is known as a *trochoid*.

Waves are generated in four fundamental ways in the open sea: (1) by changes in atmospheric pressure, (2) by the wind acting for long periods of time on the water's surface, (3) by seismic disturbances such as earthquakes, and (4) by the tidal attraction of the sun and moon. These methods of generation act at different times, in different directions, and in different amounts. When waves of different characteristics meet in a relatively small area, a complex pattern of wave motion results, caused by waves reinforcing or interfering with each other in varying amounts. This complex situation is termed *sea* and is very difficult to describe mathematically. *Wind waves* are defined as waves which are growing in height under the influence of the wind. *Swells* are defined as wind-generated waves which have left their area of generation and have advanced into regions of weaker winds or calms, and are decreasing in height. Swells assume a sort of regular undulating motion which closely approximates the so-called "ideal wave."

When waves enter shallow water, the friction caused by the bottom will tend to slow down the water particles nearest the bottom, while the upper particles continue with the faster original velocity. When the upper particles are traveling at a critical speed with respect to the bottom particles, the wave first steepens and then breaks. Waves in this condition are referred to as *breakers*.

Basically, the wave spectrum varies from short-period, wind-generated waves with periods of 3 to 30 seconds through long-period tidal waves with periods of almost 13 hours. The term *tidal wave* refers to those waves generated by the action of the sun and moon on the oceans and should not be confused with seismic-generated waves (*tsunamis*) which have been erroneously called tidal waves.

Figure 11–8. BEAUFORT FORCE 0   BEAUFORT FORCE 0–10
Wind speed less than 1 kt   Courtesy Canada,
Sea criterion: Sea like a mirror.   Atmospheric Environment Service.
Date/Time of photograph: June 5, 1960, 2340 GMT.
Height of camera above sea: 35 ft.
Waves at time of picture

	Direction (° true)	Period (sec)	Height (ft)
Sea waves	———	—	—
Swell	100	5	2

Figure 11–9. BEAUFORT FORCE 1
Wind speed 1–3 kt, mean 2 kt
Sea criterion: Ripples with the appearance of scales are formed, but without foam
                  crests.
Date/Time of photograph: May 22, 1960, 2000 GMT.
Height of camera above sea: 35 ft.
Waves at time of picture

	Direction (° true)	Period (sec)	Height (ft)
Sea waves	–––	–	–
Swell	290	10	3

Figure 11–10. BEAUFORT FORCE 2
Wind speed 4–6 kt, mean 5 kt
Sea criterion: Small wavelets, still short, but more pronounced—crests have a glassy
                  appearance and do not break.
Date/Time of photograph: May 26, 1961, 1700 GMT.
Height of camera above sea: 45 ft.
Waves at time of picture

	Direction (° true)	Period (sec)	Height (ft)
Sea waves	120	–	–
Swell	050	6	1

Figure 11–11. BEAUFORT FORCE 3
Wind speed 7–10 kt, mean 9 kt
Sea criterion: Large wavelets. Crests begin to break. Foam of a glassy appearance.
                   Perhaps scattered white horses.
Date/Time of photograph: Feb. 19, 1961, 2000 GMT.
Height of camera above sea: 45 ft.
Waves at time of picture

	Direction (° true)	Period (sec)	Height (ft)
Sea waves	–––	–	–
Swell	180	7	8

Figure 11–12. BEAUFORT FORCE 4
Wind speed 11–16 kt, mean 13 kt
Sea criterion: Small waves, becoming longer, fairly frequent white horses.
Date/Time of photograph: July 3, 1960, 2240 GMT.
Height of camera above sea: 35 ft.
Waves at time of picture

	Direction (° true)	Period (sec)	Height (ft)
Sea waves	310	5	3
Swell	–––	–	–

Figure 11–13. BEAUFORT FORCE 5
Wind speed 17–21 kt, mean 19 kt
Sea criterion: Moderate waves taking a more pronounced long form; many white horses are formed. (Chance of some spray.)
Date/Time of photograph: Apr. 7, 1961, 2315 GMT.
Height of camera above sea: 35 ft.
Waves at time of picture

	Direction (° true)	Period (sec)	Height (ft)
Sea waves	280	6	7
Swell	240	8	6

Figure 11–14. BEAUFORT FORCE 6
Wind speed 22–27 kt, mean 24 kt
Sea criterion: Large waves begin to form; the white foam crests are more extensive everywhere. (Probably some spray.)
Date/Time of photograph: Feb. 10, 1961, 2115 GMT.
Height of camera above sea: 20 ft.
Waves at time of picture

	Direction (° true)	Period (sec)	Height (ft)
Sea waves	280	6	11
Swell	———	—	——

Figure 11–15. BEAUFORT FORCE 7
Wind speed 28–33 kt, mean 30 kt
Sea criterion: Sea heaps up and white foam from breaking waves begins to be
                  blown in streaks along the direction of the wind.
Date/Time of photograph: Feb. 28, 1961, 1900 GMT.
Height of camera above sea: 45 ft.
Waves at time of picture

	Direction (° true)	Period (sec)	Height (ft)
Sea waves	300	6	13
Swell	250	9	10

Figure 11–16. BEAUFORT FORCE 8
Wind speed 34–40 kt, mean 37 kt
Sea criterion: Moderately high waves of greater length; edges of crests begin to
                  break into the spindrift. The foam is blown in well-marked streaks
                  along the direction of the wind.
Date/Time of photograph: Jan. 15, 1961, 1955 GMT.
Height of camera above sea: 35 ft.
Waves at time of picture

	Direction (° true)	Period (sec)	Height (ft)
Sea waves	260	7	18
Swell	–––	–	––

Figure 11–17. BEAUFORT FORCE 9
Wind speed 41–47 kt, mean 44 kt
Sea criterion: High waves. Dense streaks of foam along the direction of the wind. Crests of waves begin to topple, tumble, and roll over. Spray may affect visibility.
Date/Time of photograph: Jan. 17, 1961, 2130 GMT.
Height of camera above sea: 35 ft.
Waves at time of picture

	Direction (° true)	Period (sec)	Height (ft)
Sea waves	120	7	20
Swell	———	–	——

Figure 11–18. BEAUFORT FORCE 10
Wind speed 48–55 kt, mean 52 kt
Sea criterion: Very high waves with long overhanging crests. The resulting foam, in great patches, is blown in dense white streaks along the direction of the wind. On the whole, the surface of the sea takes on a white appearance. The tumbling of the sea becomes heavy and shocklike. Visibility affected.
Date/Time of photograph: Mar. 14, 1961, 2330 GMT.
Height of camera above sea: 15 ft.
Waves at time of picture

	Direction (° true)	Period (sec)	Height (ft)
Sea waves	340	9	22
Swell	———	–	——

Figure 11–19. BEAUFORT FORCE 11
Wind speed 53–63 kt, violent storm
Sea criterion: Exceptionally high waves, sea covered with white foam patches; visibility still more reduced. From *Marine Observer's Handbook*, 8th Edition, London, 1963. Courtesy of the Controller of Her Britannic Majesty's Stationery Office.

Figure 11–20. BEAUFORT FORCE 12
Wind speed 64–71 kt. hurricane
Sea criterion: Air filled with foam; sea completely white with driving spray; visibility greatly reduced. From *Marine Observer's Handbook*, 8th Edition, London, 1963. Courtesy of Her Britannic Majesty's Stationery Office.

From a mariner's point of view, the waves that most affect seamanship are those generated by the winds. It is common knowledge among seamen around the world that the longer and harder the winds blow, the higher will be the waves.

The three most important variables regarding wave relationships are the *wind* at the sea surface, the *fetch* (the stretch of water over which the wind blows), and the *duration* (the length of time that the wind has blown). Since wind, fetch, and duration vary within very wide limits, it has been necessary for meteorologists and oceanographers to develop and design special tables, charts, nomograms and slide rules to forecast all the possible variations. Although knowledge of the way in which the wind produces waves on the surface of the sea is still not complete, marine meteorologists are doing an excellent job of predicting the various oceanographic parameters. Tests made of the various methods to date indicate that wave/swell forecasts can be made with equal or greater certainty than most meteorological forecasts. The reason for this is that changes in the oceans take place much more slowly than in the atmosphere.

Table 11–1 illustrates the Beaufort Scale and enables one to estimate the Beaufort force (or wind speed) from the appearance of the sea. Beaufort number 5, for example, with winds of 17–21 knots (a knot is 1.15 mph), is a fresh breeze with an effect on the sea that produces moderate waves, which take a more pronounced long form; many white horses are formed, and there is a chance of some spray. The right-hand column of table 11–1 shows how wind reports are plotted on weather maps to the nearest five knots. Each short barb represents a speed of five knots. Each long barb represents 10 knots, and each pennant represents 50 knots. All "arrows" "fly with the wind"; they point in the direction toward which the wind is blowing.

WAVE CHARACTERISTICS AND RELATIONSHIPS

Referring to figure 11–21, a wave is described by its *length, L,* (the horizontal distance from crest to crest or trough to trough), by its *height, H,* (the vertical distance from trough to crest), and by its *period,T,* (the time interval in seconds between the appearance of two consecutive crests at a given position). A wave may be *standing* or *progressive*, but our discussion will deal with progressive waves only. In a progressive wave, if the length and energy are constant, the wave height is the same at all localities and the wave crest appears to advance with a constant speed. During one wave period T, the wave crest advances one wave length L, and the *speed* of the wave, C, is therefore defined as $C = L/T$. Waves of small height are those for which the ratio of height to length is $1/100$ or less. Waves of moderate height are those for which the ratio of H/L is from $1/100$ to $1/25$. Waves of great height are those for which the ratio of H/L is from $1/25$ to $1/7$. Theoretically, the wave form becomes unstable when the ratio of H/L exceeds $1/7$. Observational evidence, however, indicates that instability occurs at a steepness as small as $1/10$. The wave speed increases with increasing steepness (increasing values of H/L), but the increase of speed never exceeds 12 percent. The speed of a wave (in knots) can be calculated by the simple formula:

$$\text{Velocity (knots)} = 0.6 \times \frac{\text{Length (feet)}}{\text{Period (seconds)}}$$

Probably the greatest single item of controversy regarding waves is that of *maximum* wave height, because of the general tendency to underestimate small wave heights and to overestimate large wave heights. Be that as it may, the maximum wave height scientifically recorded thus far (in 1965) in the North Atlantic was 67 feet by a wave recorder aboard an ocean-station weather vessel, but this does not mean that higher waves do not occur.

Maximum Wave Height and Fetch

For a given wind velocity, the wave height becomes greater the longer the stretch of water (fetch) over which the wind has blown. Even with a very strong wind, the wave height for a given fetch does not exceed a certain maximum value.

Wave Speed and Fetch

At a given wind speed, the wave speed increases with increasing fetch.

Wave Height and Wind Speed

The height in feet of the greatest waves with high wind speeds has been observed to be about 0.8 of the wind speed in knots.

Wave Speed and Wind Speed

Although the ratio of wave speed to wind speed has been observed to vary from less than 0.1 to nearly 2.0, the average maximum wave speed apparently exceeds slightly the wind speed when the latter is less than about 25 knots, and is somewhat less than the wind speed at higher wind speeds.

Wave Height and Duration of Wind

The time required to develop waves of maximum height corresponding to a given wind increases with increasing wind speed. Observations show that with strong winds, high waves will develop in less than 12 hours.

Wave Speed and Duration of Wind

Although observational data are inadequate, it is known that for a given fetch and wind speed, the wave speed increases rapidly with time.

Wave Steepness

No well-established relationship exists between wind speed and wave steepness. This lack is probably due to the fact that wave steepness is not directly related to the wind speed, but depends upon the stage of development of the wave. The stage of development, or *age of the wave*, can be conveniently expressed by the ratio of the wave speed to the wind speed (C/U), because during the early stages of their formation, the waves are short and travel with a speed much less than that of the wind, while at later stages the wave speed may exceed the wind speed.

Decrease of Height of Swell

The height of swell decreases as the swell advances. Roughly, the swells lose one-third of their height each time they travel a distance in miles equal to their wave length in feet.

Table 11-1. The Beaufort Wind Scale, Speed Conversions and Descriptions.

Beaufort Number	Knots	Miles Per Hour	Description	Effect at sea	Wind Symbols on Weather Maps
0	0–0.9	0–0.9	Calm	Sea like a mirror.	⊙ Calm
1	1–3	1–3	Light air	Scale-like ripples form, but without foam crests.	◎ Almost Calm
2	4–6	4–7	Light breeze	Small wavelets, short but more pronounced. Crests have a glassy appearance and do not break.	5 Knots
3	7–10	8–12	Gentle breeze	Large wavelets. Crests begin to break. Foam has glassy appearance. Perhaps scattered white horses.	10 Knots
4	11–16	13–18	Moderate breeze	Small waves, becoming longer. Fairly frequent white horses.	15 Knots
5	17–21	19–24	Fresh breeze	Moderate waves, taking a more pronounced long form. Many white horses are formed. Chance of some spray.	20 Knots
6	22–27	25–31	Strong breeze	Large waves begin to form. White foam crests are more extensive everywhere. Some spray.	25 Knots
7	28–33	32–38	Moderate gale	Sea heaps up and white foam from breaking waves begins to be blown in streaks along the direction of the wind. Spindrift begins.	30 Knots

Force	Knots	mph	Description	Effects at sea	
8	34–40	39–46	Fresh gale	Moderately high waves of greater length. Edges of crests break into spindrift. Foam is blown in well-marked streaks along the direction of the wind.	35 Knots
9	41–47	47–54	Strong gale	High waves. Dense streaks of foam along the direction of the wind. Sea begins to roll. Spray may affect visibility.	45 Knots
10	48–55	55–63	Whole gale and/or Storm	Very high waves with long overhanging crests. The resulting foam in great patches is blown in dense white streaks along the direction of the wind. On the whole, the surface of the sea takes a white appearance. The rolling of the sea becomes heavy and shocklike. Visibility is affected.	50 Knots
11	56–63	64–73	Storm and/or Violent storm	Exceptionally high waves. Small- and medium-sized vessels might for a long time be lost to view behind the waves. The sea is completely covered with long white patches of foam lying along the direction of the wind. Everywhere, the edges of the wave crests are blown into froth. Visibility seriously affected.	60 Knots
12	64 or higher	74 or higher	Hurricane & Typhoon	The air is filled with foam and spray. Sea is completely white with driving spray. Visibility is very seriously affected.	75 Knots

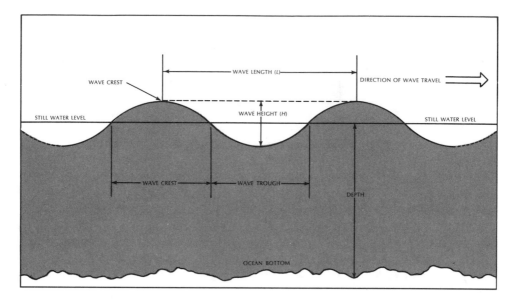

Figure 11–21. Ocean wave characteristics.

Increase of Period of Swell

Some scientists claim that the period of the swell remains unaltered when the swell advances from the generating area, whereas others claim that the period increases. The greater amount of evidence at the present time indicates that the period of the swell increases as the swell advances.

Tables 11–2 through 11–7 have been included to enable the reader to prepare his own preliminary oceanographic forecasts in conjunction with available weather data and weather maps.

Table 11–2. Probable Maximum Heights of Waves with Various Wind Speeds and Unlimited Fetch.

Wind Speed (Knots)	Wave Height (feet)
8	3
12	5
16	8
19	12
27	20
31	25
35	30
39	36
43	39
47	45
51	51

Table 11–3. Wave Heights (Feet) Produced by Different Wind Speeds Blowing for Various Lengths of Time.

Wind Speed (Knots)	Duration (Hours)						
	5	10	15	20	30	40	50
10	2	2	2	2	2	2	2
15	4	4	5	5	5	5	5
20	5	7	8	8	9	9	9
30	9	13	16	17	18	19	19
40	14	21	25	28	31	33	33
50	19	29	36	40	45	48	50
60	24	37	47	54	62	67	69

Table 11–4. Wave Heights (Feet) Produced by Different Wind Speeds Blowing Over Different Fetches.

Wind Speed (Knots)	Fetch (Nautical Miles)					
	10	50	100	300	500	1,000
10	2	2	2	2	2	2
15	3	4	5	5	5	5
20	4	7	8	9	9	9
30	6	13	16	18	19	20
40	8	18	23	30	33	34
50	10	22	30	44	47	51

Table 11–5. Minimum, Average, and Maximum Wave Heights (Feet) for the Tropical Trade-Wind Belts.

Area	Minimum	Average	Maximum
Atlantic	0	6	20
Pacific	0	10	25
Indian	3	9	16

Table 11–6. Maximum Wave Heights with Various Wind Speed and the Fetches and Durations Required to Produce Waves 75 percent as high as the Maximum with each Wind Speed.

Wind Speed (Knots)	Maximum Wave Height (Feet)	75 Percent of Maximum Height (Feet)	Fetch For 75 Percent (N.M.)	Duration for 75 Percent (Hours)
10	2	1.5	13	5
20	9	6.8	36	8
30	19	14.3	70	11
40	34	25.5	140	16
50	51	38.3	200	18

Table 11–7. Average Wave Length Compared to Wind Speed.

Average Wave Length (Feet)	Wind Speed (Knots)
52	11
124	20
261	30
383	42
827	56

SIGNIFICANT WAVES

Because of the irregular appearance of the sea surface, it is necessary to describe the waves that are present by means of some statistical term. This term should give emphasis to the higher waves because they are operationally more important than the smaller ones, although the actual number of smaller (and shorter) waves may be greater. For this reason, it is not advisable to state the mean wave height during, for example, a half-hour or one-hour period of observation, but rather to use the average height of the *highest one-third* of all observed waves.

We shall use this average and call it the *significant* wave height. This measure, as well as the mean, is not an exact measure because it depends upon the extent to which small waves have been recorded. If every ripple is counted as a wave, both the mean height and the average height of th highest one-third waves are reduced. In practice, all waves less than one foot are eliminated from consideration. Tests have indicated that the average height, rather than the mean height, of the highest one-third of all observed waves is a more consistent measure. This occurs because a casual observer tends to pay more attention to the higher waves and reports a wave height that lies closer to the significant wave height than to the mean wave height. The average height also depends less on the scope of the observations than does the mean height.

Table 11–8 shows the wave height characteristics of this measure. The significant wave height is given a relative value of 1.00. Therefore, if the significant wave height is known, the height of the maximum wave, the average height of the highest 10 percent, and the average height of the entire wave train can be computed.

By using table 11–8, it is seen, for example, that if a wave train has a significant wave height of 10 feet, the highest wave is 18.7 feet, the average of the highest 10 percent is 12.9 feet, and the mean wave height is 6.4 feet.

Table 11–8. A Comparison of Wave-Height Characteristics.

Wave Terminology	Relative Height
Significant	1.00
Average	0.64
Highest 10 Percent	1.29
Highest	1.87

ENERGY OF WAVES

A train of waves in the ocean has potential energy due to the elevation and depression of the surface from its initial state of being level, and also has kinetic energy due to the movement of every particle in a vertical circular orbit, which we discussed earlier. Theoretical reasoning shows that the amounts of potential and kinetic energy are equal and proportional to the wave length times the wave height squared (LH^2) per wave length per unit of crest length. Thus, the kinetic energy in waves is tremendous. A 4-foot, 10-second wave striking a coast expends more than 35,000 horsepower per mile of coast!

The power of waves can best be visualized by viewing the damage they cause. On the coast of Scotland, a block of cemented stone weighing 1,350 tons was broken loose and moved by waves. Five years later, the replacement pier, weighing 2,600 tons, was carried away. Engineers have measured the force of breakers along this coast of Scotland at 6,000 pounds per square foot!

Off the coast of Oregon, the roof of a lighthouse 91 feet above low water was damaged by a rock weighing 135 pounds.

An attempt has been made to harness the energy of waves along the Algerian coast. Waves are funneled through a V-shaped concrete structure into a reservoir. The water flowing out of the reservoir operates a turbine which generates power.

The energy of deep-water waves moves in their direction of travel with only half of the wave speed. This fact is responsible for the propagation of wave trains with only half the velocity of the individual waves. In any group of waves, the ones in the middle have the largest amplitude, or height, while those at the front and rear have small amplitudes which decrease to almost nothing at the edges of the group. While the group is traveling, the waves are continually overtaking the front of the wave train where they disappear, and thus the group velocity of a particular group of waves is one-half the velocity of the individual waves.

Waves lose energy when they break at the crest with the formation of "white horses," or when the wind begins to blow with a component from the opposite direction. Some of their energy is consumed in overcoming the internal friction (known as *viscocity*) which is evident between masses of water moving relative to each other. Molasses, for example, is a very viscous substance in which it is difficult to produce wave motion. The viscosity of water, however, is small, and this means that wave trains can travel great distances before they are finally dissipated.

OPTIMUM TRACK
SHIP-ROUTE FORECASTING

The shortest route between two points on our globe (as every seaman knows) is a great-circle track, from the standpoint of distance alone. But because of wind and sea conditions, it is seldom the shortest in time or the safest or most comfortable. In one year (1954), more than 6 percent of the entire world's shipping experienced heavy weather damage, and more than 3 percent were involved in collisions caused by weather and/or sea conditions.

In the early 1950s, the U.S. Navy established a ship-routing service, which is an efficient and modern version of the service provided by Lieutenant Matthew

Fontaine Maury, USN, before the Civil War, when he gathered the logs of ships and produced charts of ocean currents and winds. Maury's work resulted in saving days or weeks in the journeys of sailing vessels. Now, time savings are measured in hours and days.

The basic principles of ship routing are simultaneously simple and complex. Marine meteorologists predict wind speeds and directions for the particular areas of interest. Charts are prepared to show lines of equal wind velocity (isotachs), and these are then translated into charts which depict forecast wave heights, as shown in figure 11–22. By use of these charts, maximum attainable and safe speed can be computed for any type of ship or vessel. A ship using this service maintains communications with the organization supplying the ship-routing service and receives a daily course to be steered. Many millions of dollars are now saved each year in faster transits, less fuel consumed, minimizing storm damage, and so forth.

Although optimum track ship-route forecasting was developed by the U.S. Navy, civilian private forecasters and organizations now furnish the same service to commercial ships and private individuals.

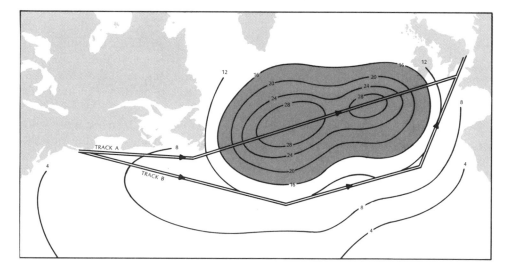

Figure 11–22. Optimum track ship-routing across the Atlantic. Track A is the great circle route from New York to Southampton, England. Track B is the optimum ship track recommended by marine meteorologists in this weather/oceanographic situation. Contours are predicted wave heights in feet. Shading indicates areas of waves in excess of 16 feet. Although Track A is the shortest distance between New York and Southampton, ships following it would encounter 28-foot waves, would have to slow their speed of advance considerably, would take longer to arrive at their destination, and would probably sustain storm damage either to the ship or cargo or both. On the recommended Track B, maximum wave heights would be only 12 feet, and the voyage would be made in less time and with no storm damage.

DETERMINATION OF STORM DIRECTION
AND DISTANCE BY OBSERVING SWELL

If the height and period of the swell reaching you are observed, it is possible to determine *approximate* values of the distance of the storm (or more accurately, the end of the wave-generating area) from which the swell came; the travel time of the swell (the length of time it took the swell to travel from the storm to your position); and the wind speed in the storm (or generating area) that created the swell. This is most important information to have, in order that a sound decision may be made regarding storm evasion, other courses of action, and so forth.

The values that can be derived from figure 11–23, it must be re-emphasized, are only approximate, because the height and period of the swell depend also upon the relation between wave and wind speeds (C/U) at the end of the fetch (the generating area of the waves). For our purpose, the ratio is assumed to vary in accordance with an assumed relationship between wind speed and duration (the length of time the wind has blown). In figure 11–23, two specific relationships between wind speed and duration are shown in the inset of the diagram. In the upper and lower parts of the diagram, corresponding values of decay distance, travel time, and wind speed are shown.

Choice of the better relationship between wind speed and duration must be based on a knowledge of weather situations that prevail in the area under consideration, and on the common experience that high winds are usually of relatively short duration, while weaker winds may blow for a long period of time. For some combinations of observed swell height and period, only one of the two parts of figure 11–23 will apply.

As mentioned earlier, one must realize that the results obtained from figure 11–23 are approximate, since the wind speed and duration relationship may vary considerably from one weather situation to another. A lack of complete knowledge regarding the changes caused by following or opposing winds, and inaccuracies in the observations of the swell height and period will also introduce errors, though small. The values of distance and travel time obtained from figure 11–23 are more accurate than values of wind speed. We will try two examples to ensure our knowledge of how to use the diagram.

CASE 1: As we cruise along at greatly reduced speed, we are subjected to the effects of a 20-foot swell with a 12-second period, coming from a direction of 070 degrees true. What is the distance of the storm (or generating area) from our vessel? What is the bearing of the storm from our position? How long did it take the swells to reach our position? What was the wind speed that generated the swells now causing us this discomfort?

Referring to figure 11–23, in the upper diagram, the dot marked X is plotted at the intersection of the horizontal line labelled *20 feet*, and the vertical line labelled *12 seconds*. The three sets of curves on the diagram provide us with three of the answers to this problem. Interpolating between the three sets of curves, (distance from which the swell comes, travel time, and wind speed in the generating area or storm), we obtain the following answers:

1. The storm (or generating area) is 440 nautical miles away from our position.
2. At the time the swell was generated, the storm (or generating area) was on a

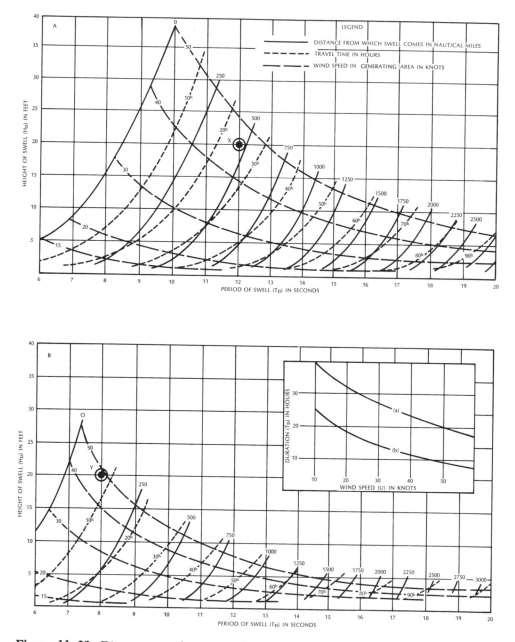

Figure 11–23. Diagrams to determine the distance from which a swell comes, the travel time of the swell, and the wind speed in the generating area, as functions of the observed height and period of the swell.

bearing of 070 degrees true from our position. (But remember, it will have moved in the time interval between generation and arrival at our position).
3. It took the swells 24 hours to reach our position.
4. Wind speeds in the storm (or generating area) which created the swells were 46 knots.

CASE 2: In the same sort of circumstances mentioned in CASE 1, we observe a 20-foot swell with an 8-second period, coming from a direction of 135 degrees true. Again, what is the distance of the storm (or generating area) from our vessel? What is the bearing of the storm from our position? How long did it take the swells to reach our position? What was the wind speed which generated the swells?

Referring to figure 11–23, again, in the lower diagram, the dot marked Y plotted at the intersection of the horizontal line labelled *20 feet*, and the vertical line labelled 8 *seconds* represents the observed conditions of the swell. Again, interpolating between the three sets of curves in the lower diagram, we obtain the following answers:
1. The storm (or generating area) is 120 nautical miles distant from our position.
2. At the time the swell was generated, the storm (or generating area) was on a bearing of 135 degrees true from our position.
3. It took the swells eight hours to reach our position.
4. Wind speeds in the storm (or generating area) that created the swells were 47 knots.

OTHER BENEFITS FROM A KNOWLEDGE OF WAVES

At the present time, a great deal remains to be learned regarding the generation, travel, and decay of ocean waves and swell. However, we do know quite a bit about these phenomena at the present state of the art. A knowledge of the height and other characteristics of waves is of considerable practical value for a variety of purposes, including the design and behavior of vessels at sea; the design and orientation of harbors and the construction of breakwaters; problems of coast erosion and silting; the discharging of ships in open anchorages; and naval amphibious and carrier operations. For example, the behavior of an individual vessel or small craft at sea is governed to a considerable extent by the period of her roll and pitch in various conditions of loading in relation to the period of the waves she encounters, and her longitudinal and transverse strength calculations must inevitably take into account similar factors.

As mentioned in chapter 8, in tropical and subtropical waters, the arrival of a rather gentle swell (if from an unusual direction) may well be the first warning of the approach of a dangerous hurricane or typhoon. Similarly, wave data have value in areas where direct observation of storms are not available because of the limited amount of shipping in the area. From time to time, public investigations and inquiries are held into shipping losses where the destruction of life and property was serious, and in such cases, evidence provided by actual observations of wave or swell height from vessels in the vicinity has always proved to be of great value.

12
simple forecasting methods and rules

"When the wind backs
  And the weatherglass falls,
  Then be on your guard
  Against gales and squalls."

                    —Source unknown

In order that professional and amateur weathermen alike may understand, follow, and predict the weather, detailed observations must be taken *at the same time* over extensive regions, such as the entire northern hemisphere, and exchanged internationally. When an observer on the east coast of the United States makes a weather observation at 7:00 a.m. Eastern Standard Time, an observer on the Pacific Coast makes his at 4:00 a.m. Pacific Standard Time, and an observer in London, England makes his weather observation at noon Greenwich Time. These simultaneous observations provide the *synoptic* weather picture, the state of the earth's atmosphere at that particular instant of time. Surface observations are made around the world at least every six hours at 0000, 0600, 1200, and 1800 Greenwich Time, and upper-air observations are made at least every 12 hours at 0000 and 1200 Greenwich Time. More frequent observations are made when required or in special circumstances.

This wealth of information must be presented visually in such form that the users of weather charts can determine the current situation almost at a glance, perform the various analyses, and prepare appropriate forecasts. The pictorial presentation of weather data on the different charts serves this purpose.

HOW TO READ WEATHER MAPS

The information at each station of the weather networks of the world (and of ships at sea), whether surface or upper-air observations, is arranged around a circle drawn at the location of the station (and current position of the ships at sea). Analyses are performed depending upon the particular purpose of the chart.

The following weather information is plotted for each station and ship on a surface weather chart, and each item is always plotted in exactly the same position relative to the station/ship circle: wind direction and speed, pressure, temperature, dew point, visibility, ceiling, current weather (rain, snow, fog, etc.), the amount and types of clouds and their heights, pressure changes in the past three hours, weather in the past six hours, and the amount and type of precipitation. This tremendous amount of information for each station (and ship) is shown by symbols and numbers around the circle in a small space easily covered by a dime. Figures 12–1 and 12–2 are examples of land-station and ship weather plots, respectively, as they appear on surface weather maps. The relative position of each weather element around the circle should be memorized by the reader. Figure 12–3 contains all the weather symbols and notations used by weathermen in all countries of the world by international agreement. The reader is not expected to memorize all these symbols—though meteorologists must—but should keep the chart handy for reference when working with weather maps.

Daily weather maps of various types are available from a variety of sources—the NOAA Weather Service, most daily newspapers, private weather consultants, industrial meteorological organizations, and so forth. Figure 12–4 includes examples of NOAA Weather Service daily weather maps which are available at a nominal cost from the Superintendent of Documents, Government Printing Office, Washington, D.C. 20402. The surface weather map presents station data and the analysis for 0700 EST. The tracks of well-defined low-pressure areas are indicated by chains of arrows (when applicable). The locations of these centers at

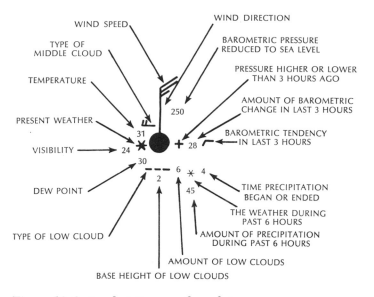

Figure 12–1. Land-station weather plot.

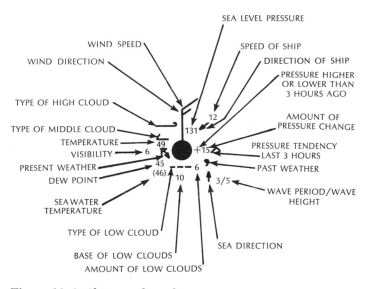

Figure 12–2. Ship weather plot.

the times 6, 12, and 18 hours preceding map time are indicated by small black squares enclosing white crosses. Areas of precipitation are indicated by shading. The weather reports printed on these maps are only a small fraction of those that are included in the operational weather maps, and on which analyses are based. Occasional apparent discrepancies between the printed station data and the analyses result from those station reports that cannot be included in the pub-

00 Cloud development NOT observed or NOT observable during past hour.§	01 Clouds generally dissolving or becoming less developed during past hour.§	02 State of sky on the whole unchanged during past hour.§	03 Clouds generally forming or developing during past hour.§	04 Visibility reduced by smoke.	05 Haze.	06 Widespread dust in suspension in the air, NOT raised by wind, at time of observation.	07 Dust or sand raised by wind, at time of ob.	08 Well developed dust devil(s) within past hr.
10 Light fog.	11 Patches of shallow fog at station, NOT deeper than 6 feet on land.	12 More or less continuous shallow fog at station, NOT deeper than 6 feet on land.	13 Lightning visible, no thunder heard.	14 Precipitation within sight, but NOT reaching the ground.	15 Precipitation within sight, reaching the ground, but distant from station.	16 Precipitation within sight, reaching the ground, near to but NOT at station.	17 Thunder heard, but no precipitation at the station.	18 Squall(s) within sight during past hour.
20 Drizzle (NOT freezing and NOT falling as showers) during past hour, but NOT at time of ob.	21 Rain (NOT freezing and NOT falling as showers) during past hr., but NOT at time of ob.	22 Snow (NOT falling as showers) during past hr., but NOT at time of ob.	23 Rain and snow (NOT falling as showers) during past hour, but NOT at time of observation.	24 Freezing drizzle or freezing rain (NOT falling as showers) during past hour, but NOT at time of observation.	25 Showers of rain during past hour, but NOT at time of observation.	26 Showers of snow, or of rain and snow, during past hour, but NOT at time of observation.	27 Showers of hail, or of hail and rain, during past hour, but NOT at time of observation.	28 Fog during past hour, but NOT at time of ob.
30 Slight or moderate duststorm or sandstorm, has decreased during past hour.	31 Slight or moderate dustorm or sandstorm, no appreciable change during past hour.	32 Slight or moderate dustorm or sandstorm, has increased during past hour.	33 Severe duststorm or sandstorm, has decreased during past hr.	34 Severe duststorm or sandstorm, no appreciable change during past hour.	35 Severe duststorm or sandstorm, has increased during past hour.	36 Slight or moderate drifting snow, generally low.	37 Heavy drifting snow, generally low.	38 Slight or moderate drifting snow, generally high.
40 Fog at distance at time of ob., but NOT at station during past hour.	41 Fog in patches.	42 Fog, sky discernible, has become thinner during past hour.	43 Fog, sky NOT discernible, has become thinner during past hour.	44 Fog, sky discernible, no appreciable change during past hour.	45 Fog, sky NOT discernible, no appreciable change during past hour.	46 Fog, sky discernible, has begun or become thicker during past hr.	47 Fog, sky NOT discernible, has begun or become thicker during past hour.	48 Fog, depositing rime, sky discernible.
50 Intermittent drizzle (NOT freezing) slight at time of observation.	51 Continuous drizzle (NOT freezing) slight at time of observation.	52 Intermittent drizzle (NOT freezing) moderate at time of ob.	53 Continuous drizzle (NOT freezing), moderate at time of ob.	54 Intermittent drizzle (NOT freezing), thick at time of observation.	55 Continuous drizzle (NOT freezing), thick at time of observation.	56 Slight freezing drizzle.	57 Moderate or thick freezing drizzle.	58 Drizzle and rain, slight.
60 Intermittent rain (NOT freezing), slight at time of observation.	61 Continuous rain (NOT freezing), slight at time of observation.	62 Intermittent rain (NOT freezing), moderate at time of ob.	63 Continuous rain (NOT freezing), moderate at time of ob.	64 Intermittent rain (NOT freezing), heavy at time of observation.	65 Continuous rain (NOT freezing), heavy at time of observation.	66 Slight freezing rain.	67 Moderate or heavy freezing rain.	68 Rain or drizzle and snow, slight.
70 Intermittent fall of snow flakes, slight at time of observation.	71 Continuous fall of snowflakes, slight at time of observation.	72 Intermittent fall of snow flakes, moderate at time of observation.	73 Continuous fall of snowflakes, moderate at time of observation.	74 Intermittent fall of snow flakes, heavy at time of observation.	75 Continuos fall of snowflakes, heavy at time of observation.	76 Ice needles (with or without fog).	77 Granular snow (with or without fog).	78 Isolated starlike snow crystals (with or without fog).
80 Slight rain shower(s).	81 Moderate or heavy rain shower(s).	82 Violent rain shower(s).	83 Slight shower(s) of rain and snow mixed.	84 Moderate or heavy shower(s) of rain and snow mixed.	85 Slight snow shower(s).	86 Moderate or heavy snow shower(s).	87 Slight shower(s) of soft or small hail with or without rain or and snow mixed.	88 Moderate or heavy shower(s) of soft or small hail with or without rain or and snow mixed.
90 Moderate or heavy shower(s) of hail††, with or without rain or rain and snow mixed, not associated with thunder.	91 Slight rain at time of ob.; thunderstorm during past hour, but NOT at time of observation.	92 Moderate or heavy rain at time of ob.; thunderstorm during past hour, but NOT at time of observation.	93 Slight snow mixed or hail at time of observa.; thunderstorm during past hour, but not at time of observation.	94 Mod. or heavy snow or rain and snow mixed or hail at time of ob.; thunderstorm during past hour, but NOT at time of observation.	95 Slight or mod. thunderstorm without hail, but with rain and/or snow at time of ob.	96 Slight or mod. thunderstorm, with hail at time of observation.	97 Heavy thunderstorm, without hail, but with rain and/or snow at time of observation.	98 Thunderstorm combined with duststorm or sandstorm at time of ob.

SYMBOLIC FORM OF SYNOPTIC CODE: iii Nddff VVwwW PPPTT $N_hC_LhC_MC_H$ $T_dT_d$app 7RRR₁s 8N$_s$C$_h$h$_s$

Figure 12–3. Weather code figures and symbols with their meanings.

lished maps because of lack of space. The symbols of fronts, pressure systems, air masses, and so forth, are the same as discussed in previous chapters. Note how the isobars are V-shaped (they "kink") at the fronts, as previously discussed.

The 500-*Millibar Chart* presents the height contours and isotherms of the 500-millibar surface at 0700 EST. The 500-millibar heights usually range from 15,800 to 19,400 feet. The height contours are shown as continuous lines and are labeled in feet above sea level. The isotherms are shown as dashed lines and are

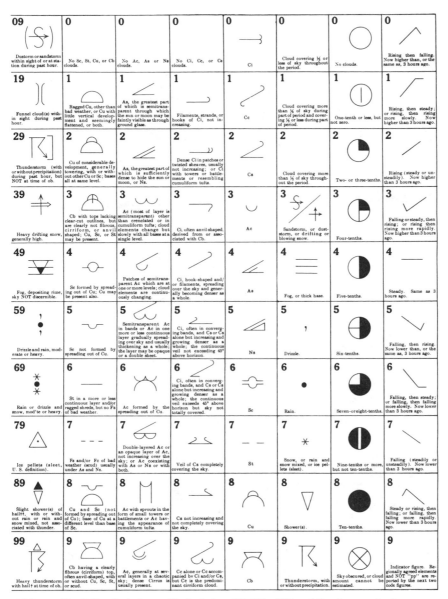

ww	CL	CM	CH	C	W	N	a
09 Dust storm or sandstorm within sight of or at station during past hour.	**0** No Sc, St, Cu, or Cb clouds.	**0** No Ac, As or Ns clouds.	**0** No Ci, Cc, or Cs clouds.	**0** Ci	**0** Cloud covering ½ or less of sky throughout the period.	**0** No clouds.	**0** Rising then falling. Now higher than, or the same as, 3 hours ago.
19 Funnel cloud(s) within sight during past hour.	**1** Ragged Cu, other than bad weather, or Cu with little vertical development and seemingly flattened, or both.	**1** As, the greatest part of which is semitransparent through which the sun or moon may be faintly visible as through ground glass.	**1** Filaments, strands, or hooks of Ci, not increasing.	**1** Cc	**1** Cloud covering more than ½ of sky during part of period and covering ½ or less during part of period.	**1** One-tenth or less, but not zero.	**1** Rising, then steady; or rising, then rising more slowly. Now higher than 3 hours ago.
29 Thunderstorm (with or without precipitation) during past hour, but NOT at time of ob.	**2** Cu of considerable development, generally towering, with or without Cu or Sc; bases all at same level.	**2** As, the greatest part of which is sufficiently dense to hide the sun or moon, or Ns.	**2** Dense Ci in patches or twisted sheaves, usually not increasing; or Ci with towers or battlements or resembling cumuliform tufts.	**2** Cs	**2** Cloud covering more than ½ sky throughout the period.	**2** Two- or three-tenths.	**2** Rising (steady or unsteadily). Now higher than 3 hours ago.
39 Heavy drifting snow, generally high.	**3** Cb with tops lacking clear-cut outlines, but are clearly not fibrous, cirriform, or anvil shaped; Cu, Sc, or St may be present.	**3** Ac (most of layer is semitransparent) other than crenelated or in cumuliform tufts; cloud elements change but slowly with all bases at a single level.	**3** Ci, often anvil-shaped, derived from or associated with Cb.	**3** Ac	**3** Sandstorm, or dust-storm, or drifting or blowing snow.	**3** Four-tenths.	**3** Falling or steady, then rising; or rising then rising more rapidly. Now higher than 3 hours ago.
49 Fog, depositing rime, sky NOT discernible.	**4** Sc formed by spreading out of Cu; Cu may be present also.	**4** Patches of semitransparent Ac which are at one or more levels; cloud elements are continuously changing.	**4** Ci, hook-shaped and/or filaments, spreading over the sky and generally becoming denser as a whole.	**4** As	**4** Fog, or thick haze.	**4** Five-tenths.	**4** Steady. Same as 3 hours ago.
59 Drizzle and rain, moderate or heavy.	**5** Sc not formed by spreading out of Cu.	**5** Semitransparent Ac in bands or Ac in one or more or less continuous layer gradually spreading over sky and usually thickening as a whole; the layer may be opaque or a double sheet.	**5** Ci, often in converging bands, and Cs or Cs alone but increasing and growing denser as a whole; the continuous veil not exceeding 45° above horizon.	**5** Ns	**5** Drizzle.	**5** Six-tenths.	**5** Falling, then rising. Now lower than, or the same as, 3 hours ago.
69 Rain or drizzle and snow, mod'te or heavy.	**6** St in a more or less continuous layer and/or ragged shreds, but no Fs of bad weather.	**6** Ac formed by the spreading out of Cu.	**6** Ci, often in converging bands, and Cs alone but increasing and growing denser as a whole; the continuous veil exceeds 45° above horizon but sky not totally covered.	**6** Sc	**6** Rain.	**6** Seven-or-eight-tenths.	**6** Falling, then steady; or falling, then falling more slowly. Now lower than 3 hours ago.
79 Ice pellets (sleet, U. S. definition).	**7** Fs and/or Fc of bad weather (scud) usually under As and Ns.	**7** Double-layered Ac or an opaque layer of Ac, not increasing over the sky; or Ac coexisting with As or Ns or with both.	**7** Veil of Cs completely covering the sky.	**7** St	**7** Snow, or rain and snow mixed, or ice pellets (sleet).	**7** Nine-tenths or more, but not ten-tenths.	**7** Falling (steadily or unsteadily). Now lower than 3 hours ago.
89 Slight shower(s) of hail†, with or without rain or rain and snow mixed, not associated with thunder.	**8** Cu and Sc (not formed by spreading out of Cu); base of Cu at a different level than base of Sc.	**8** Ac with sprouts in the form of small towers or battlements or Ac having the appearance of cumuliform tufts.	**8** Cs not increasing and not completely covering the sky.	**8** Cu	**8** Shower(s).	**8** Ten-tenths.	**8** Steady or rising, then falling; or falling, then falling more rapidly. Now lower than 3 hours ago.
99 Heavy thunderstorm with hail† at time of ob.	**9** Cb having a clearly fibrous (cirriform) top, often anvil-shaped, with or without Cu, Sc, St, or scud.	**9** Ac, generally at several layers in a chaotic sky; dense Cirrus is usually present.	**9** Cc accompanied by Ci and/or Cs, but Cc is the predominant cirriform cloud.	**9** Cb	**9** Thunderstorm, with or without precipitation.	**9** Sky obscured, or cloud amount cannot be estimated.	**9** Indicator figure. Regionally agreed elements and NOT "pp" are reported by the next two code figures.

$9S_pS_ps_ps_p$ $1d_wd_wP_wH_w$ $2h_{85}h_{85}h_{85}a_3$ $3R_{24}R_{24}R_{24}R_{24}$ $4T_xT_xT_nT_n$ (Additional Plain Language Data)

labeled in degrees Centigrade. The arrows show the wind direction and speed at the 500-millibar level.

The *Highest and Lowest Temperature Chart* presents the maximum and minimum values for the 24-hour period ending at 0100 EST. The names of the reporting points can be obtained from the surface weather map. The maximum temperature is plotted above the station location, and the minimum temperature is plotted below this point.

WEDNESDAY, DECEMBER 4, 1968

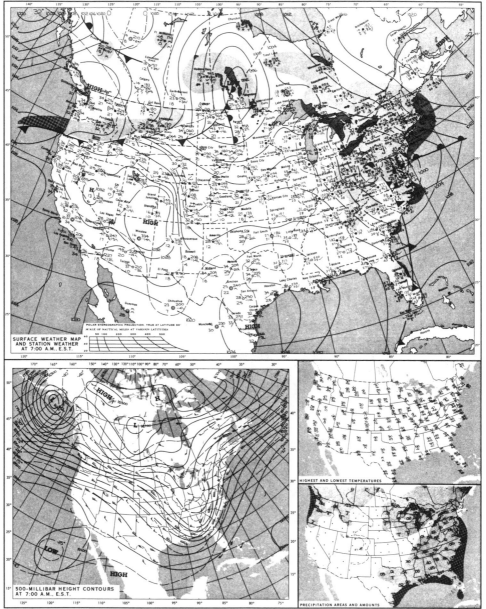

Figure 12–4. NOAA National Weather Service daily weather maps.

The *Precipitation Areas and Amounts Chart* indicates by means of shading the areas that had precipitation during the 24 hours ending at 0100 EST. Amounts in inches to the nearest hundredth of an inch are for the same period. Incomplete totals are underlined. A *T* indicates only a trace of precipitation. Dashed lines show the depth of snow on the ground in inches as of 0700 EST of the previous day.

Figure 12–5 is an example of an American forecast (prognostic) surface weather map that appears in *The New York Times*, and figure 12–6 illustrates its British counterpart that appears in *The Guardian*.

Although newspaper daily weather maps are not as detailed as the official charts of the NOAA National Weather Service or foreign governmental meteorological organizations, they are timely, readily accessible, inexpensive, and can be used to great advantage by the professional mariner and weekend sailor with some knowledge of the weather science. At sea, weather maps obtained by radio facsimile serve the same purpose.

BEING YOUR OWN WEATHER FORECASTER

According to most meteorologists, there is no sharp distinction between analyzing a weather map and making a weather forecast. When one analyzes a weather map, one studies the physical properties of the air masses; the structures, positions, and past movements of the weather fronts and pressure centers, and so forth, all in relation to space and time. Consequently, as one analyzes a weather map, one is automatically thinking of what will happen in the future. Before making a weather prediction, there are many questions which must be answered relative to the forecast period. Some of these are: what will be the direction and speed of *movement* of the pressure systems, the weather fronts, and so forth? What will be the changes in *intensity* of the pressure systems, the weather fronts, and so forth? What will be the changes in the *physical properties* of the air masses? The most simple method of forecasting is concerned primarily with "extrapolating" future weather conditions on the basis of present and past conditions. In reality, forecasts by professional meteorologists are prepared on the basis of many scientific principles, the assistance of high-speed computers and weather satellites, and many years of experience. Although the reader cannot claim to be a professional meteorologist at this point, he should have absorbed enough information to be able to prepare a reasonable, short-term forecast for a period of 6–12 hours into the future.

In simple short-range forecasting, using a single weather map, the assumption is usually made that pressure systems (highs and lows), weather fronts and their associated weather, and so forth, will behave in the same way for the next 12 hours as they behaved in the previous 12 hours. In other words, a front (or a pressure system) which moved at an average speed of 25 knots for the past 12 hours will continue to move at that same average speed and in the same direction, with the same associated weather, for the next 12 hours. A front or pressure system which was stationary for the past 12 hours will continue to remain stationary for the next 12 hours, and so on.

Let's try our hand at what *could* be an uncomplicated situation and an easy forecast. Referring to the simplified weather map of figure 12–7, note the continuous precipitation ahead of the warm front and the rainshowers behind the cold front. It is your job to prepare a forecast for Boston, Massachusetts. At 1000 a.m. EST, the cold front located 150 nautical miles to the west of Boston is moving eastward at 25 knots. What is your forecast?

At the reported speed of 25 knots, the cold front will travel the 150-nautical-mile distance to Boston in 6 hours, and will arrive there at 1600 p.m. EST. Based

Weather Reports and Forecast

Summary

Clear skies and near seasonable temperatures are expected today throughout most of the Northeast. Except for a few showers over the Florida peninsula, it will be fair from the Eastern Seaboard to the western edges of the Plains States; cooler conditions will prevail along the Atlantic Coast. It will be warmer however, from the Mississippi Valley to the Rockies. Rain and snow-showers will be scattered in the Rockies and intermountain region, while along the Pacific Coast, cooler air and rain are forecast.

Clouds covered the eastern third of the country yesterday including the Gulf Coast, except for Metropolitan New York, portions of the Middle Atlantic States, and the Florida peninsula; snow-showers were scattered across the northern Appalachians, lower lake region and upper Ohio Valley, while heavy showers and thunder-showers were scattered in the South Atlantic States. It was unseasonably warm along the Eastern Seaboard, and unseasonably cool from the lower Mississippi Valley to southern California. Skies were mostly fair from the Mississippi Valley to the Rockies, although clouds covered the northwestern quarter of the country. Rain along the coast of the Pacific Northwest changed to snow as it spread inland to the northern intermountain region; heavy snow was reported in most of Idaho, northern Nevada and the mountains and high plateaus of northern California.

Forecast

National Weather Service (As of 5 P.M.)
NEW YORK CITY—Sunny and seasonable today, high in the mid-40's; northwesterly winds 10 to 15 miles an hour today, becoming variable less than 10 miles an hour tonight; clear tonight, low in the mid-30's. Sunny and milder tomorrow. Precipitation probability near zero through tonight.

NORTHERN NEW JERSEY AND ROCKLAND AND WESTCHESTER COUNTIES—Sunny and seasonable today, high in the mid-40's; clear tonight, low near 20 inland valleys, and in the mid-30's along the coast. Sunny and milder tomorrow.

LONG ISLAND AND LONG ISLAND SOUND—Sunny and seasonable today, high in the mid-40's; northwesterly winds 10 to 15 miles an hour today, becoming variable less than 10 miles an hour tonight; clear tonight, low in the 20's east, and in the low to mid-30's elsewhere. Sunny and milder tomorrow. Visibility on few five miles or better through tonight.

SOUTHERN NEW JERSEY AND EASTERN PENNSYLVANIA—Mostly sunny today, high in the mid-30's northwest and in the 40's elsewhere; fair tonight, low in the 20's north and in the low

to mid-30's south. Mostly sunny and milder tomorrow.
WESTERN PENNSYLVANIA—Sunny today, high in the 30's; clear tonight, low in the teens to mid-20's. Mostly sunny and milder tomorrow.
WESTERN NEW YORK—Cloudy with chance of a few flurries today, high in the mid-30's; variably cloudy and not as cold tonight and tomorrow, low tonight in the low to mid-20's.
CONNECTICUT, RHODE ISLAND AND MASSACHUSETTS—Sunny today, high in the low to mid-40's; clear tonight, low in the 20's. Fair and continued seasonable tomorrow.
NEW HAMPSHIRE AND MAINE—Sunny to partly sunny today, high near 30 northern Maine and in the 40's south; fair and continued cold tonight and tomorrow, low tonight 5 above zero northern Maine and in the teens to low 20's elsewhere.

Extended Forecast

(Saturday through Monday)
METROPOLITAN NEW YORK, LONG ISLAND AND NORTHERN NEW JERSEY—Chance of rain Saturday and Sunday; fair Monday. Daytime highs will average near 50, while overnight lows average near 40.
CONNECTICUT, RHODE ISLAND AND MASSACHUSETTS—Chance of rain Saturday and Sunday; daytime highs will average in the 40's, while overnight lows average in the 30's. Fair Monday; daytime highs will average in the low 30's to low 40's, while overnight lows average in the 20's.

Yesterday's Records

Eastern Standard Time

	Temy.	Hum.	Winds		Bar.
1 A.M.	43	93	SE	3	29.69
2 A.M.	42	96	SE	6	29.66
3 A.M.	42	96	SE	6	29.63
4 A.M.	43	99	SE	6	29.60
5 A.M.	43	96	NW	5	29.60
6 A.M.	43	96	NW	5	29.59
7 A.M.	42	96	NW	5	29.60
8 A.M.	43	96	NW	6	29.62
9 A.M.	43	96	NW	8	29.63
10 A.M.	45	89	NW	9	29.63
11 A.M.	48	74	NW	16	29.63
Noon	48	66	NW	18	29.63
1 P.M.	50	59	NW	14	29.62
2 P.M.	48	63	NW	12	29.61
3 P.M.	46	63	NW	17	29.62
4 P.M.	46	60	NW	14	29.64
5 P.M.	44	58	NW	10	29.66
6 P.M.	43	62	NW	12	29.68
7 P.M.	42	65	NW	9	29.70
8 P.M.	42	62	6W	11	29.71

Temperature Data

(19-hour period ended 7 P.M.)
Lowest, 44 at 1:10 A.M.
Highest, 51 at 12:20 P.M.
Mean, 46.
Normal on this date, 34.
Departure from normal, +12.
Departure this month, —5.
Departure this year, +160.
Lowest this date last year, 33.
Highest this date last year, 49.
Mean this date last year, 40.
Lowest temperature this date, 1 in 1936.
Highest temperature this date, 63 in 1961.
Lowest mean this date, 10 in 1936.
Highest mean this date, 52 in 1961.
Degree day yesterday*, 19.
Degree days since Sept. 1, 3,155.
Normal since Sept. 1, 3,312.
Total last season to this date, 3,080.
*A degree day (for heating) indicates the number of degrees the mean temperature falls below 65 degrees. The American Society of Heating, Refrigeration and Air-Conditioning Engineers has designated 65 degrees as the point below which heating is required.

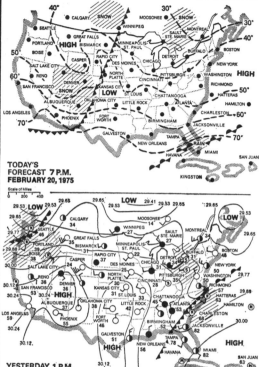

TODAY'S FORECAST 7 P.M. FEBRUARY 20, 1975

YESTERDAY 1 P.M. FEBRUARY 19, 1975

Figure beside Station Circle is temperature.
Cold front: a boundary between cold air and warmer air, under which the colder air pushes like a wedge, usually south and east.
Warm front: a boundary between warm air and a retreating wedge of colder air over which the warm air is forced as it advances, usually north and east.
Occluded front: a line along which warm air was lifted by opposing wedges of cold air, often causing precipitation.
Shaded areas indicate precipitation.
Dash lines show forecast afternoon maximum temperatures.
Isobars are lines (solid black) of equal barometric pressure (in inches), forming air-flow patterns.
Winds are counterclockwise toward the center of low-pressure systems, clockwise outward from high-pressure areas. Pressure systems usually move east.

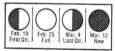

National Weather Service Map, N.O.A.A.
U.S. Department of Commerce

Precipitation Data

(24-hours ended 7 P.M.)
Twelve hours ended 7 A.M., .12.
Twelve hours ended 7 P.M., .0.
Total this month to date, 1.72.
Total since January 1, 6.48.
Normal this month, 2.92.
Days with precipitation this date, 42 since 1869.
Least amount this month, 46 in 1895.
Greatest amount this month, 6.87 in 1869.
1869.

Sun and Moon

The sun rises today at 6:43 A.M.; sets at 5:35 P.M., and will rise tomorrow at 6:41 A.M.
The moon rises today at 11:45 A.M.; sets tomorrow at 2:59 A.M., and will rise tomorrow at 12:46 P.M.

Feb. 19 First Qtr. Feb. 25 Full Mar. 4 Last Qtr. Mar. 12 New

Planets

New York City
(Tomorrow, Eastern Standard Time)
Venus—rises 7:48 A.M.; sets 7:42 P.M.
Mars—rises 4:50 A.M.; sets 2:11 P.M.
Jupiter—rises 7:40 A.M.; sets 7:23 P.M.
Saturn—rises 1:16 P.M.; sets 4:17 A.M.
Planets rise in the East and set in the West, reaching their highest point on the north-south meridian, midway between their times of rising and setting.

Figure 12–5. Daily weather map in a United States newspaper. Courtesy of the *New York Times*.

THE WEATHER: Bright spells and showers

A complex area of low pressure over the British Isles will move away to NW. Scotland and Northern Ireland will have a good deal of cloud, but there will be some bright intervals. Most parts will have rain or snow, and there are likely to be some longer outbreaks of rain or snow in N Scotland. N England will have sunny intervals and wintry showers, but some places may be rather misty at first. Wales and the rest of England will have bright spells and showers. They will be most frequent in W districts, and will be of sleet and snow at times, especially over high ground. More general rain or snow may spread into SW later in the day. Temperatures will be generally below normal with frost in many places early and late.

London area, SE and Cent S England, East Anglia, Midlands: Sunny intervals and scattered showers. Sleet at times. Wind W, moderate. Max. temp. 5C (41F).

E England, Cent N and NE England: Misty at first. Sunny intervals and scattered wintry showers later. Wind W. moderate. Max. 4C (39F).

Channel Islands, SW England, S Wales: Sunny intervals and wintry showers. Perhaps more general rain or snow later. Wind W. becoming SW, moderate. Max. 6C (43F).

N Wales, NW England, Lake District, Isle of Man, SW Scotland, Glasgow area, Argyll, N Ireland: Bright intervals and wintry showers. Wind W. moderate or fresh. Max 5C (41F).

Borders, Edinburgh, E Scotland, Aberdeen area, Cent Highlands: Bright intervals and scattered wintry showers. Perhaps misty at first. Wind N moderate. Max 5C (41F).

Moray Firth area, Caithness, NW Scotland, Orkney, Shetland: Rather cloudy. Occasional rain, sleet, or snow. Perhaps some bright intervals. Wind S to SW. moderate. Max 4C (39F).

Outlook: Continuing rather cold with rain at times, and some snow over high ground. Frost in many places at night.

SEA PASSAGES

S North Sea, Strait of Dover, English Channel (E): Moderate. St George's Channel, Irish Sea: Moderate to rough.

THE SATELLITES

The figures give in order: Time of visibility; where rising; maximum elevation and direction of setting. An asterisk indicates entering or leaving eclipse.
Pageos A. 18.17-18.41 N 85 SSE S and 21.19-21.33 NNW 25 WNW NNW.

LIGHTING-UP TIMES

Birmingham .... 5 55 p.m. to 8 40 a.m.
Bristol .... 6 03 p.m. to 8 37 a.m.
London .... 5 53 p.m. to 8 27 a.m.
Nottingham .... 5 51 p.m. to 8 30 a.m.

HIGH-TIDE TABLE

London Bridge 1 29 a.m. .... 1 59 p.m.
Dover ...... 11 04 a.m. .... 11 35 p.m.

SUN RISES .... 8 59 a.m
SUN SETS ...... 5 23 p.m.
MOON RISES .... 9 06 a.m.
MOON SETS ...... 3 59 p.m.
MOON: New Moon tomorrow.

LONDON READINGS

7 p.m. Wednesday to 7 a.m. yesterday: Temp. min. 2C (36F); rainfall .16in, 7 a.m. to 7 p.m. yesterday; Temp. max. 7C (45F); rainfall, nil; sunshine 1.1 hours.

Around Britain

Reports for the 24 hours ended 6 p.m. yesterday:

	Sunshine hrs.	Rain in.	Max. temp. C	F	Weather (day)
EAST COAST					
Scarboro ...		.23	4	40	Rain p.m.
Bridlingtn ...	.78	4	39	Rain
Lowestoft ...		.47	5	41	Rain
Clacton ...	1.5	.64	4	40	Cloudy
Whitstable	1.3	.54	6	42	Cloudy
Herne Bay	1.6	.51	5	41	Cloudy
SOUTH COAST					
Folkestone	1.8	.17	5	41	Cloudy
Hastings	2.0	.25	6	42	Cloudy
Eastbourne .	2.9	.16	6	42	Sunny intls
Brighton ...	3.4	.39	6	43	Sunny intls
Worthing ..	3.9	.31	6	43	Sunny intls
Littlehptn	2.7	.40	1	44	Sunny intls
Bognor Reg.	4.4	.20	6	43	Sunny prds
Southsea ..	2.4	.50	6	43	Sunny intls
Sandown ...	2.2	.5..	6	43	Sunny intls
Shanklin ...	2.7	.66	5	41	Sunny intls
Ventnor ...	3.0	.50	6	42	Shwr
Bournemth	1.3	.11	6	42	Rain
Poole	2.0	.21	6	43	Shwrs
Swanage	3.5	.09	6	43	Sunny intls
Weymouth			8	47	Drizzle
Exmouth ..	1.8	.33	6	42	Sleet shwrs
Teignmtn	2.2	.34	6	43	Shwrs
Torquay ...	5.0	.57	5	41	Sunny prds
Falmouth .	6.3	.21	7	44	Sunny prds
Penzance ...	5.8	.21	8	46	Gales. sunny prds
Jersey	5.3	.50	8	47	Sunny prds
WEST COAST					
Douglas ..		.37	5	41	Rain. gales
Morecambe .			2	36	Fog
Blackpool		.04	2	36	Sleet
Southport ..		.05	3	38	Fog, snow
Colwyn Bay		.27	6	42	Rain
Llandudno ..		.18	6	42	Rain
Anglesey	0.1	.47	6	43	Rain
Weston-s-M.		.16	4	40	Cloudy
Ilfracombe ..		.17	7	45	Rain. gales
Newquay ..	3.3	.13	6	43	Snny ins. gls
Scilly Isles	5.9	.05	8	46	Sunny prds
INLAND					
Ross-on-Wy	1.7	.07	5	41	Cloudy
SCOTLAND					
Lerwick ..	0.7	.28	6	43	Hail
Wick		.28	5	41	Rain
Stornoway .		.15	6	43	Rain
Kinloss ...	0.1	.21	5	41	Rain
Dyce		.45	4	39	Rain
Tiree	0.1	.35	6	43	Rain
Leuchars ..	0.3	.20	5	41	Rain, snow
Glasgow ...		.06	5	41	Rain
Eskdalemuir		.01	3	37	Snow. rain
NORTHERN IRELAND					
Belfast	2.1	.05	5	41	Showers

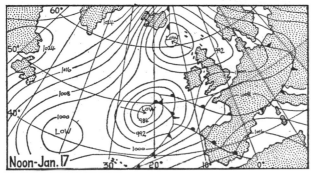

Noon-Jan. 17

Around the world
Lunch-time reports

		C	F				C	F
Algiers ..	C	16	61		Lisbon ..	F	15	59
Amsterdm	C	6	43		Locarno .	Sl	1	34
Athens ..	S	16	61		London ..	R	6	43
Barcelona	S	12	54		Luxembrg	C	4	39
Beirut ..	C	18	64		Madrid ..	F	10	50
Belfast ..	C	4	39		Majorca..	S	14	57
Belgrade .	S	11	52		Malaga ..	C	18	64
Berlin ..	C	2	36		Malta ...	C	15	59
Birminghm	R	3	37		Miami ..	Dr	3	37
Bristol ..	F	4	39		Moscow ..	C	-6	21
Brussels..	C	5	41		Munich ..	S	-2	28
Budapest.	C	3	37		Naples ..	R	11	52
Cardiff ..	C	4	39		Nice	S	11	52
Cologne .	C	7	45		Nicosia ..	C	13	55
Copenhgn	Fo	2	36		Oslo	C	1	54
Dublin ..	C	1	34		Paris	R	4	39
Edinburgh	C	5	41		Prague ..	C	-1	30
Faro	C	16	61		Reykjavik	F	-6	21
Florence .	R	8	46		Rome	Th	9	48
Frankfurt	C	2	36		Ronaldswy	R	5	41
Geneva ..	C	3	37		Stockholm	Dr	2	36
Gibraltar .	F	17	63		Tel-Aviv .	C	17	63
Guernsey .	F	6	43		Toronto ..	C	-3	27
Helsinki .	C	-1	30		Tunis	C	16	61
Innsbruck	C	0	32		Venice ..	C	8	46
Istanbul .	C	12	54		Vienna ..	C	1	34
Jersey ..	F	7	45		Warsaw ..	C	3	37
Las Palmas	C	19	66		Zurich ..	F	3	37

C, cloudy; Dr, drizzle; S, sunny; R, rain; Fo, fog; Sl, sleet; Th, thunderstorm.

Snow reports

	L	U	Piste	Off piste	Wthr	F
Andermatt	52	86	Good	Powder	Cloud	28
Davos	24	38	Good	Powder	Snow	36
Grindewald	14	38	Good	Powder	Fine	31
Gstaad ...	10	30	Good	Powder	Snow	32
Lenzerheide	41	60	Good	Varied	Cloud	32
Murren ...	27	44	Good	Powder	Snow	30
Pontresina	45	80	Good	Powder	Fair	32
Sauze d'Oulx	18	60	Good	Powder	Fine	26
Verbier ..	18	36	Good	Varied	Snow	40
Zermatt ..	42	66	Good	Powder	Cloud	32

In the above reports, supplied by representatives of the Ski Club of Great Britain L refers to lower slopes and U to upper slopes. The following reports have ben received from other sources.

Depth in

	L	U	Piste	Wthr	F
ITALY					
Bardonecchia ..	24	92	Good	Fine	28
Bormio	24	48	Good	Snow	29
Canazei	16	80	Good	Fine	23
Cervinia	48	100	Good	Fine	34
Claviere	38	60	Good	Fine	19
Cortina	73	120	Good	Fine	—
Courmayeur ..	22	92	Good	Fine	28
Macugnaca	20	120	Good	Cloud	29
Madesimo	60	160	Good	Cloud	23
Madonna di Campiglio	23	60	Good	Cloud	29
San Martino ..	12	44	Good	Fine	29
Selva	28	52	Good	Fine	29
GERMANY					
Berchtesgaden ..	10	36	Poor	Cloud	26
Garmisch	10	22	Poor	Fine	21
Hindelang	16	40	Poor	Cloud	23
Kleinwalsertal ..	22	40	Good	Fine	23
Mittenwald ...	8	52	Poor	Fine	26
Oberammergau .	10	28	Poor	Fine	26
Oberjoch	16	40	Poor	Cloud	21
Oberstaufen ...	22	32	Poor	Fine	24
Oberstdorf	16	32	Poor	Fine	15
AUSTRIA					
Badgastein	10	35	Fair	Cloud	30
Berwang	9	35	Good	Cloud	30
Galtur	16	28	Good	Cloud	26
Mayrhofen	10	16	Good	Cloud	25
Obergurgl	16	35	Good	Cloud	35
Saalbach	8	12	Fair	Cloud	35
Schruns	21	32	Fair	Cloud	26
Solden	25	46	Good	Cloud	23
NORWAY		U	Piste	Wthr	
Finse		22	Good	Cloud	
Gol		28	Good	Snow	
Geilo		28	Good	Cloud	
Voss		32	Good	Fair	
Lillehammer ..		33	Good	Cloud	
Oslo		39	Good	Cloud	

SCOTLAND

Cairngorms: Main runs: General snow cover. New snow on a firm base. Lower slopes: General snow cover. New snow on a firm base. Maximum runs: 1,600ft. Access roads, slight snow.
Glenshee: Main runs: All runs complete. Powdery snow on a firm base. Lower slopes: Ample nursery areas Powder snow on a firm base. Maximum runs: 1,000ft. Access roads: Clear.
Glencoe: Main runs: Some runs complete, others broken. Hard packed snow. Lower slopes: Ample nursery areas. Hard packed snow. Maximum: 1,100ft. Access roads: Clear.
Forecast: Snow showers, but also bright periods. Moderate W winds. Freezing level 500ft.

Figure 12–6. Daily weather map and other information in a British newspaper. Courtesy of Guardian Newspapers, Ltd.

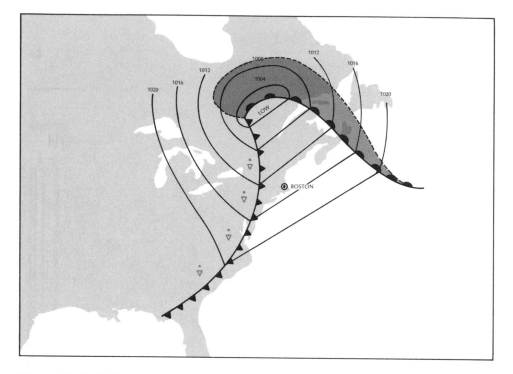

Figure 12–7. Making your own short-range weather forecast. It is now 1000 EST. At this time, the cold front is 150 nautical miles to the west of Boston and is moving eastward at 25 knots. What is your forecast for Boston?

on what you have learned thus far, your forecast for Boston should be: "Cold front passage accompanied by rainshowers at 1600 p.m. EST. Surface winds shifting from SW to NW during frontal passage. Cooler tonight and tomorrow."

But sometimes, what appears to be a rather simple weather situation turns out to be something quite different. This is why we occasionally shovel several inches of "partly cloudy" out of our driveways and off our sidewalks during the winter months. In our simple case of figure 12–7, for example, as the cold front sweeps across the eastern seaboard into the Atlantic, a brand-new, secondary low-pressure system *could* develop to the south of Cape Hatteras, North Carolina (a favorite breeding area of secondary cyclones) and travel rapidly northeastward, paralleling the coastline. In this case, a typical "Nor'easter" would soon envelop the Boston area. A trained and well-experienced forecaster can accurately estimate and predict these developments, but on occasion, they are caught, too. A beginner in the weather business would not be likely to spot such a development in advance.

The easiest way to determine the future movement of pressure systems is to do a freehand extrapolation of their paths. This method has been appropriately named the "path method" by Professor S. Petterssen. For this method, a series of weather maps (rather than a single weather map) is required. Figure 12–8 il-

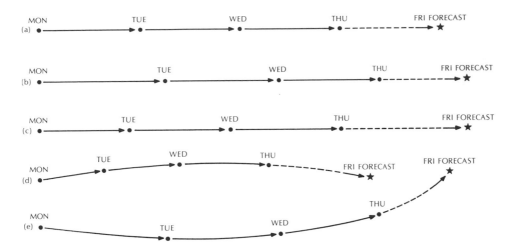

Figure 12–8. The "path method" of forecasting. Looking from left to right, the "dots" show the consecutive positions of a pressure system from Monday through Thursday. The "stars" indicate the forecast position for Friday. (a) Direction of movement in straight line; speed of movement constant. (b) Direction of movement in straight line; speed of movement decreasing. (c) Direction of movement in straight line; speed of movement increasing. (d) Direction of movement curving to right; speed of movement increasing. (e) Direction of movement curving to left; speed of movement decreasing.

lustrates only 5 of many typical paths. Figure 12–8a shows 4 consecutive positions (Mon. thru Thurs.) and the extrapolated (forecast) position for Friday, when the direction of movement is in a straight line, and the speed of movement is constant. Figure 12–8e shows 4 consecutive positions (Mon. thru Thurs.) and the extrapolated (forecast) position for Friday, when the direction of movement curves to the left, and the speed of movement decreases. Basically, the "path method" of forecasting involves the extrapolation into the future of established trends—whether they be speed changes, intensity changes, directional-movement changes, some of them, or all of them. Well, let's try our hand at Petterssen's "path method," using a very simplified series of weather maps.

Figure 12–9 shows a 1008-millibar low-pressure system centered over the state of Colorado on a Monday, with a cold front extending to the southwest out of the low, and a warm front extending to the southeast. Rainshowers are occurring behind the cold front, and a steady rain is falling ahead of the warm front.

In figure 12–10, on Tuesday, the low has deepened by 4 millibars, with the central pressure now having a value of 1004 millibars. The low, itself, is now centered over extreme western Illinois, having traveled eastward at 30 knots, covering a distance of 720 nautical miles in the past 24 hours. The cold front is beginning to overtake the warm front, as evidenced by the warm sector (the area between the cold front and the warm front, south of the low center) becoming more narrow. The precipitation area ahead of the warm front becomes wider.

In figure 12–11, on Wednesday, the low has deepened by another 4 millibars, and the central pressure is now 1,000 millibars. The low is now centered over the

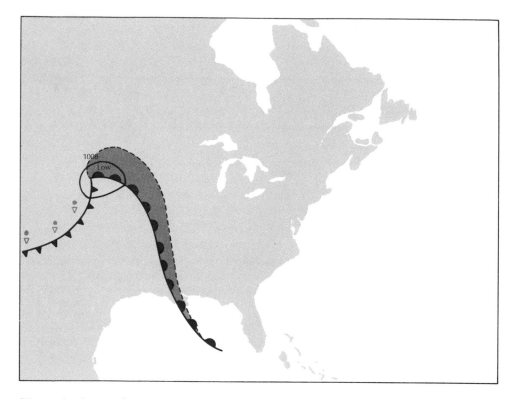

Figure 12–9. Simplified surface weather map for Monday.

extreme southwestern portion of the state of New York, having traveled east-northeastward at 25 knots, covering a distance of 600 nautical miles in the past 24 hours. The cold front is now rapidly overtaking the warm front, and the frontal wave is just beginning to occlude. Steady precipitation is spreading farther in advance of the warm front.

In figure 12–12, on Thursday, the low-pressure system has deepened by an additional 8 millibars, and the central pressure has fallen to 992 millibars. The low is now centered in extreme northern Maine, having curved slightly from an east-northeast track to a northeast track and having decelerated further to an average speed of 20 knots in the past 24 hours, covering a distance of 480 nautical miles in the last 24-hour period. The cold air mass has now overtaken the warm air mass and formed an occluded front that extends from the center of the low to a position of latitude 40 degrees north, longitude 65 degrees west. The precipitation is much the same as before, except that the steady rain now extends over the top of the occlusion.

What would you expect Friday's weather map to look like? How would you draw your forecast (prognostic) chart for tomorrow (Friday)?

One of the simplest and best methods of doing this is to take a blank map and place it over the Monday, Tuesday, Wednesday, and Thursday maps in turn, tracing the frontal positions and the lowest encircling isobar of the lows (for each

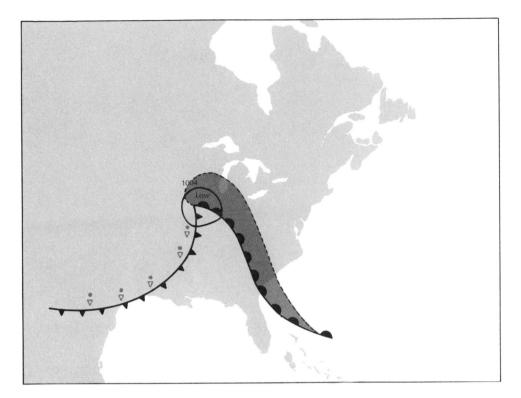

Figure 12–10. Simplified surface weather map for Tuesday.

day) onto the blank map. Note the changes in value of the central pressure of the low. Next, draw arrows connecting the centers of the lows for each day. Note that the connecting arrows become a little shorter each 24-hour period (the low is decelerating) and that the arrows point a little more toward the north each day (the system is curving toward the north in its trajectory).

To predict the Friday position of the center of the low, continue the smooth curve (which connects the central points of the low on Monday through Thursday) toward the northeast, curving it slightly more to the north. During the past three days, the speed of forward movement of the low has decelerated by 5 knots during each 24-hour period. Consequently, a continued deceleration of 5 knots should be forecast for the next 24-hour period. Thus, the average speed of movement (forward) of the low over the next 24-hour period should be 15 knots. This means that the low will travel a distance of 360 nautical miles between Thursday and Friday. Place a mark along the smooth curve (trajectory of the low) previously drawn, at a distance of 360 nautical miles from the center of the low on Thursday. This is your predicted position of the center of the low on Friday—24 hours into the future. Also, since the central pressure of the low dropped by 8 millibars in the past 24 hours, it can be expected to drop another 8 millibars in the next 24 hours, and a central pressure of 984 millibars should be predicted for Friday.

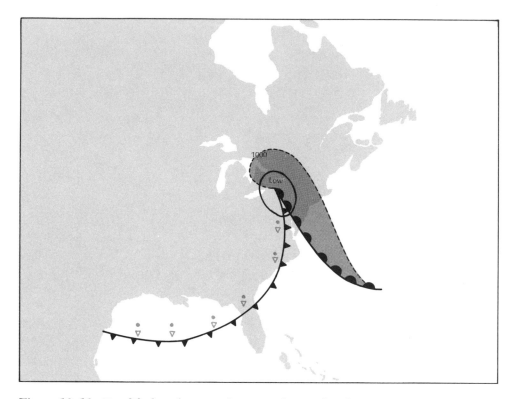

Figure 12–11. Simplified surface weather map for Wednesday.

Next, in exactly the same manner, draw 24-hour movement arrows for both the cold and the warm fronts. Also, be sure to draw in the precipitation shield, noting how it changes each 24-hour period. Do not hesitate to draw several arrows along each front in order to obtain the changes (if any) in orientation. Finally, connect the frontal-arrow tips for each front to smooth in the frontal positions. This is your forecast map for tomorrow (Friday).

Having gone through this drill systematically and carefully, your work chart and prognostic weather map should resemble very closely figure 12–13. If this is not the case, you'd better try again.

USING UPPER-AIR CHARTS IN FORECASTING

Figures 12–14 and 12–15 are NOAA National Weather Service (formerly ESSA Weather Bureau) upper-air charts. They are the 0700 EST 500-millibar charts (average height 18,280 feet) for 13 and 14 December 1968, respectively. The chart notations are as explained in the previous section of this chapter. Note that on December 13th, a closed low-pressure cell (called a *cut-off low*) existed at this pressure level over southeastern Minnesota. On the following day, December 14th, the cut-off low had disappeared, and only a deep trough of low pressure, oriented NNE–SSW, extended from extreme northern Labrador to eastern Loui-

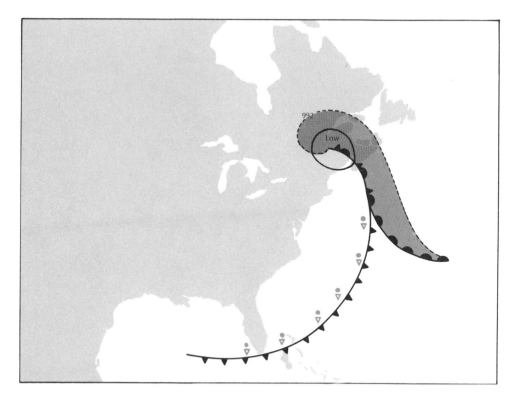

Figure 12–12. Simplified surface weather map for Thursday.

siana. These two types of upper-air patterns are most important to remember, as we shall see shortly.

The direction and speed of movement (and other factors) of surface pressure systems and fronts are influenced to a great degree by the air currents (and other conditions) at higher levels in the atmosphere. When the speed of upper-air currents is slow and these currents have great curvature, the movement of surface pressure and/or frontal systems is also slow, as shown in figure 12–16. The 24-hour movement of the surface system, represented by the dot, is only 240 nautical miles. When the speed of upper-air currents is high and these currents have little or no curvature, the movement of surface pressure and/or frontal systems is also high, as shown in figure 12–17. The 24-hour movement of the surface system in this case, represented by the dot, is 720 nautical miles. From this, it is readily apparent that some knowledge and understanding of upper-air flow patterns is essential for accurate surface weather forecasting.

WEATHER SIGNS AND SOME GENERAL RULES FOR FORECASTING

Certainly, one should not hope to be a weather expert and infallible forecaster after reading a few books and articles on the subject and preparing a few prognostic charts. Meteorologists who have been involved in all aspects of the weather

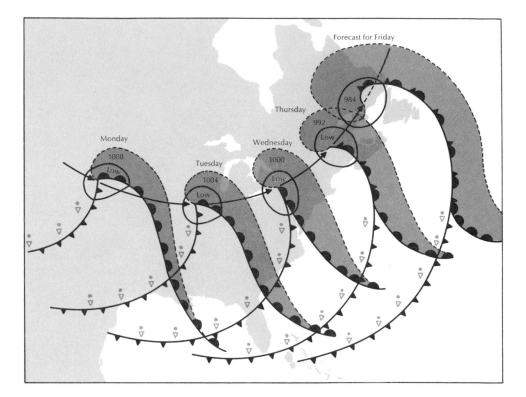

Figure 12–13. Work chart and prognostic surface weather map for Friday.

business for decades are the first to admit that they *still* learn something new about the weather with each day that passes.

There are many proverbs, rules, slogans, and signs pertaining to weather forecasting. Some are complementary, some are contradictory, and some are even correct—at least part of the time. Realizing this, we should heed carefully the counsel given over 2,300 years ago by the Greek poet, Aratus, who wrote in his *On Weather Signs:* "Make light of none of the warnings. It is a good rule to look for sign confirming sign. When two point the same way, forecast with hope. When three point the same way, forecast with confidence."

Weather signs *do* have prediction value if you have some sort of weather map available to you, and if you understand the atmospheric conditions which the signs indicate. If you can explain to yourself and to others—and you *should* be able to do so now—what the following weather signs and rules-of-thumb mean (in more-or-less scientific terminology), then they will be of definite use to you. The weather signs and forecasting rules already discussed in previous chapters will not be repeated here.

Signs

Fair weather will generally continue when:
    Summer fog clears off before noon.

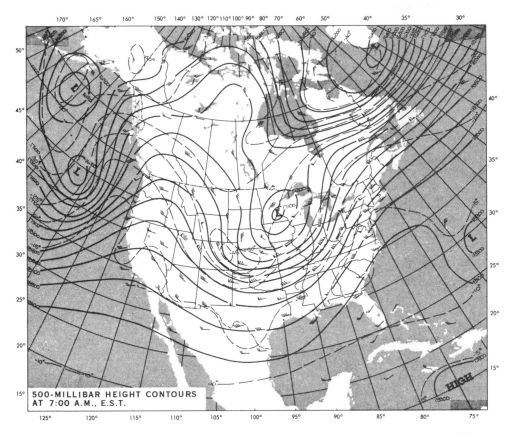

Figure 12–14. NOAA National Weather Service 500-millibar chart for 13 December.

Cloud bases along mountains increase in height.
Clouds tend to decrease in number.
The wind blows gently from west to northwest.
The temperature is "normal" for the time of year.
The barometer is steady or rising slowly.
The setting sun looks like a "ball of fire" and the sky is clear.
The moon shines brightly and the wind is light.
There is a heavy dew or frost at night.

Weather will generally change for the worse when:
Cirrus clouds change to cirrostratus and lower and thicken.
Rapidly moving clouds increase in number and lower in height.
Clouds move in different directions at different heights.
Clouds are moving from between NNE through east to south, and the wind speed increases with time.
Altocumulus or altostratus clouds darken the western horizon and the barometer begins to fall rapidly.
The wind shifts to the south or east. The greatest change occurs when the wind shifts from north through east to south.

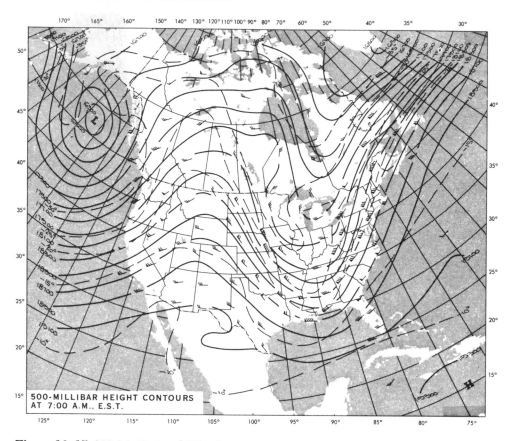

Figure 12–15. NOAA National Weather Service 500-millibar chart for 14 December.

The wind blows strongly in the early morning.
The temperature rises abnormally in the winter.
The temperature is far above or below "normal" for that time of year.
The barometer falls steadily.
There is a downpour at night.
A cold front, warm front, or occluded front approaches.

Weather will generally clear when:
Cloud bases increase in height.
A cloudy sky shows signs of clearing.
The wind shifts to a westerly direction. The greatest change occurs when the
wind shifts from east through south to west.
The barometer rises rapidly.
A cold front has passed three to six hours ago.

Rain or snow will generally occur:
When a cold, warm, or occluded front approaches.
In about 20 to 40 hours after the first cirrus-type clouds are noted to thicken
and lower.

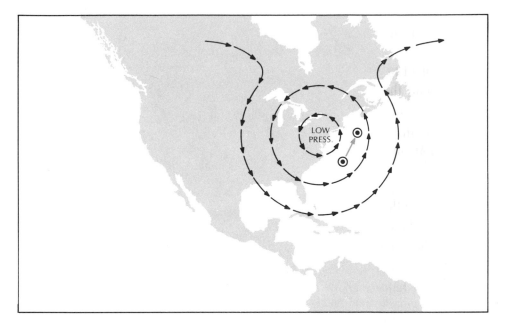

Figure 12–16. When the speed of upper-air currents over surface systems is slow and these currents have large curvature, the movement of surface pressure and/or frontal systems is also slow. The 24-hour movement of a surface pressure system, represented by the "dots," is only 240 nautical miles.

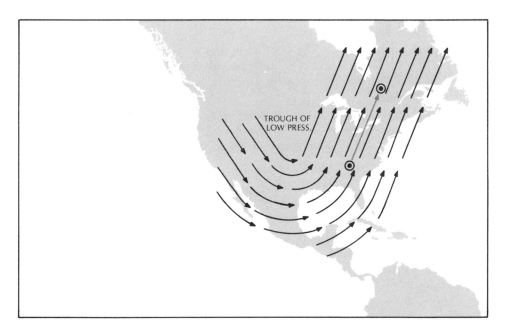

Figure 12–17. When the speed of upper-air currents over surface systems is high and these currents have little or no curvature, the movement of surface pressure and/or frontal systems is also high. The 24-hour movement of a surface pressure system, represented by the "dots," is 720 nautical miles. This is 480 nautical miles farther away than the case of figure 12–16.

In about 14 to 26 hours after cirrostratus clouds are noted and there is a halo around the sun or moon.

Within about six to eight hours when the morning temperature is unusually high, the air is humid, and cumulus clouds are observed to be building.

Within about an hour in the afternoon when there is static on the radio and swelling cumulus clouds are observed.

When the sky is dark and threatening to the west.

When a southerly wind increases in speed and the clouds above are moving from the west.

When the wind—especially a north wind—backs (shifts in a counterclockwise manner from north to west to south).

When the barometer falls steadily.

Temperature will generally fall when:
The wind shifts into the north or northwest.
The wind continues to blow from the north or northwest.
The night sky is clear and the wind is light.
The barometer rises steadily in winter.
A cold front has passed.

Temperature will generally rise when:
The sky is overcast and there is a moderate southerly wind at night.
The sky is clear during the day and there is a moderate southerly wind.
The wind shifts from the west or northwest to south.
A warm front has passed.

Fog will generally form when:
The sky is clear at sunset, the wind is light, and the air is humid.
Warm rain is falling through cold air ahead of a warm front.
There is a large temperature difference betweeen relatively warm water and much colder air above it.
There is a sustained flow of warm, moist air northward (from the south) over a colder surface (either land or water).
Remember to watch carefully the temperature-dew point spread.

Rules of Thumb

Wind Direction	Sea-Level Pressure Millibars (Inches)	General Forecast
SW to NW	1019.3 (30.10) to 1022.7 (30.20) and steady	Fair, with little temperature change, for 1 to 2 days.
SW to NW	1019.3 (30.10) to 1022.7 (30.20) rising rapidly	Fair, followed within 2 days by rain.

Rules of Thumb (*Cont.*)

Wind Direction	Sea-Level Pressure Millibars (Inches)	General Forecast
SW to NW	1022.7 (30.20) or higher and steady	Continued fair with little temperature change.
SW to NW	1022.7 (30.20) or higher falling slowly	Fair for 2 days with slowly rising temperature.
S to SE	1019.3 (30.10) to 1022.7 (30.20) falling slowly	Rain within 24 hours.
S to SE	1019.3 (30.10) to 1022.7 (30.20) falling rapidly	Increasing winds and rain within 12 to 24 hours.
SE to NE	1019.3 (30.10) to 1022.7 (30.20) falling slowly	Increasing winds and rain within 12 to 18 hours.
S to SW	1015.9 (30.00) or below rising slowly	Clearing within a few hours. Then fair for several days.
S to E	1009.1 (29.80) or below falling rapidly	Severe storm within a few hours. Then clearing within 24 hours—followed by colder in winter.
SE to NE	1019.3 (30.10) to 1022.7 (30.20) falling rapidly	Increasing winds and rain within 12 hours.
SE to NE	1015.9 (30.00) or below falling slowly	Rain will continue 1 to 3 days, perhaps even longer.
SE to NE	1015.9 (30.00) or below falling rapidly	Rain with high winds in a few hours. Clearing within 36 hours—becoming colder in winter.
E to NE	1019.3 (30.10) or higher falling slowly	In summer, with light winds, rain may not fall for 2 to 3 days. In winter, rain within 24 hours.
E to NE	1019.3 (30.10) or higher falling rapidly	In summer, rain probably within 12 to 24 hours. In winter, rain or snow within 12 hours and increasing winds.
E to N	1009.1 (29.80) or below falling rapidly	Severe storm (typical Nor'easter) in a few hours. Heavy rains or snowstorm. Followed by a cold wave in winter.
Hauling to W	1009.1 (29.80) or below rising rapidly	End of the storm. Followed by clearing and colder.

Figure 12–18. The advent of NOAA's SMS/GOES geostationary weather satellites orbiting over the equator at a height of 22,300 miles will make the work of weather forecasters somewhat easier and more accurate. Upper photo taken on 20 Feb. 1975. Lower photo shows how conditions have changed in three days, on 23 Feb. 1975. (Courtesy of NOAA National Environmental Satellite Service).

CONCLUSION

In this modern era of power-driven vessels with radio, radar, electronic aids, and satellite communications, there is an inevitable tendency for professional mariners and weekend sailors to be somewhat less weatherwise than their predecessors in square-riggers. But the practical value to all seamen of knowing something about the weather and the sea has not lessened with the passage of time. In fact, in this scientific, technological, and competitive age, the majority of seafarers agree that it has *increased*. The wind and the sea are still as powerful as ever. Hurricanes, typhoons, fog, ice, and adverse currents, are still factors to be respected and avoided. Many a modern maritime casualty has had weather and sea conditions or both as a contributory cause.

appendices

APPENDIX A. HANDY CONVERSION TABLES

Table A-1. Meters to Feet

Meters	Feet	Meters	Feet	Meters	Feet
1	3.3	10	32.8	100	328.1
2	6.6	20	65.6	200	656.2
3	9.8	30	98.4	300	984.3
4	13.1	40	131.2	400	1312.3
5	16.4	50	164.0	500	1640.4
6	19.7	60	196.9	600	1968.5
7	23.0	70	229.7	700	2296.6
8	26.2	80	262.5	800	2624.7
9	29.5	90	295.3	900	2952.8
				1000	3280.8

Table A-2. Nautical Miles to Statute Miles and Kilometers.

Nautical Miles	Statute Miles	Kilometers	Nautical Miles	Statute Miles	Kilometers
1	1.15	1.85	10	11.52	18.53
2	2.30	3.71	20	23.03	37.07
3	3.46	5.56	30	34.56	55.60
4	4.61	7.41	40	46.06	74.13
5	5.76	9.27	50	57.58	92.66
6	6.91	11.12	60	69.09	111.20
7	8.06	12.97	70	80.61	129.73
8	9.21	14.83	80	92.12	148.26
9	10.36	16.68	90	103.64	166.79

NOTE: For larger distances, merely shift the decimal point.

Table A–3. Knots to Miles Per Hour and Meters Per Second.

Knots	Miles Per Hour	Meters Per Second	Knots	Miles Per Hour	Meters Per Second
1	1.2	0.5	70	80.6	36.0
2	2.3	1.0	80	92.1	41.2
3	3.5	1.5	90	103.6	46.3
4	4.6	2.1	100	115.2	51.5
5	5.8	2.6	110	126.7	56.6
6	6.9	3.1	120	138.2	61.8
7	8.1	3.6	130	149.7	66.9
8	9.2	4.1	140	161.2	72.1
9	10.4	4.6	150	172.7	77.2
10	11.5	5.1	160	184.2	82.4
20	23.0	10.3	170	195.8	87.5
30	34.5	15.4	180	207.3	92.7
40	46.1	20.6	190	218.8	97.8
50	57.6	25.7	200	230.3	103.0
60	69.1	30.9			

Table A–4. Speed and Distance Table.

Knots	Nautical Miles Per Day	Nautical Miles Per Week	Knots	Nautical Miles Per Day	Nautical Miles Per Week
3	72	504	14.5	348	2,436
3.5	84	588	15	360	2,520
4	96	672	15.5	372	2,604
4.5	108	756	16	384	2,688
5	120	840	16.5	396	2,772
5.5	132	924	17	408	2,856
6	144	1,008	17.5	420	2,940
6.5	156	1,092	18	432	3,024
7	168	1,176	18.5	444	3,108
7.5	180	1,260	19	456	3,192
8	192	1,344	19.5	468	3,276
8.5	204	1,428	20	480	3,360
9	216	1,512	20.5	492	3,444
9.5	228	1,596	21	504	3,528
10	240	1,680	21.5	516	3,612
10.5	252	1,764	22	528	3,696
11	264	1,848	22.5	540	3,780
11.5	276	1,932	23	552	3,864
12	288	2,016	23.5	564	3,948
12.5	300	2,100	24	576	4,032
13	312	2,184	24.5	588	4,116
13.5	324	2,268	25	600	4,200
14	336	2,352			

Table A–5. Visible Distance to the Horizon (Based on Height of Observer's Eye).

Height (Feet)	Visible Distance			Dip of Horizon (Min./Sec.)
	Nautical Miles	Statute Miles	Kilometers	
5	2.57	2.96	4.77	2′ 10″
10	3.63	4.18	6.73	3 04
15	4.45	5.12	8.24	3 45
20	5.13	5.92	9.53	4 20
25	5.74	6.61	10.64	4 51
30	6.29	7.24	11.65	5 18
40	7.26	8.36	13.46	6 07
50	8.12	9.35	15.05	6 51
70	9.61	11.06	17.81	8 06
100	11.48	13.22	21.28	9 41
150	14.06	16.19	26.07	11 52
200	16.23	18.69	30.09	13 42
300	19.88	22.90	36.87	16 44
400	22.96	26.44	42.57	19 32
500	25.67	29.56	47.59	21 39
1,000	36.30	41.80	67.30	30 37

Table A–6. Altitude and Atmosphere Table.

Height (Thousands of feet)	Pressure Millibars	Pounds Per Sq. Ft.	Density ($x\ 10^4$) Lbs. Per Cu. Ft.	Temperature (Degrees C.)
0	1,013	2,116	765	+15.0
1	977	2,040	743	+12.5
2	942	1,967	721	+11.0
3	908	1,897	700	+ 8.5
4	875	1,828	679	+ 7.0
5	843	1,760	659	+ 5.1
6	812	1,696	639	+ 3.1
7	781	1,632	620	+ 1.1
8	752	1,572	601	− 0.9
9	724	1,513	583	− 2.8
10	697	1,455	565	− 4.8
12	644	1,346	530	− 8.8
14	595	1,243	497	−12.7
16	549	1,147	466	−16.7
18	506	1,057	436	−20.6
20	466	972	407	−22.4
25	376	785	343	−34.5
30	301	628	286	−44.4
35	238	498	237	−54.3
40	188	392	188	−56.5
45	147	308	148	−56.5
50	116	242	116	−56.5
55	91	196	92	−56.5
60	72	150	72	−56.5
65	59	118	57	−56.5
80	27.5	57	27.6	−46.3
100	10.8	22.6	10.1	−40.3
120	4.6	9.6	3.96	−21.6
140	2.06	4.3	1.66	− 3.3
160	0.97	2.0	0.75	+ 9.7
180	0.46	0.97	0.36	+ 3.1
200	0.21	0.44	0.18	−18.2

Table A–7. Temperatures: Degrees Centigrade to Fahrenheit.

Degrees Centigrade	0	1	2	3	4	5	6	7	8	9
+40	104.0	105.8	107.6	109.4	111.2	113.0	114.8	116.6	118.4	120.2
+30	86.0	87.8	89.6	91.4	93.2	95.0	96.8	98.6	100.4	102.2
+20	68.0	69.8	71.6	73.4	75.2	77.0	78.8	80.6	82.4	84.2
+10	50.0	51.8	53.6	55.4	57.2	59.0	60.8	62.6	64.4	66.2
+ 0	32.0	33.8	35.6	37.4	39.2	41.0	42.8	44.6	46.4	48.2
− 0	32.0	30.2	28.4	26.6	24.8	23.0	21.2	19.4	17.6	15.8
−10	14.0	12.2	10.4	8.6	6.8	5.0	3.2	1.4	− 0.4	− 2.2
−20	− 4.0	− 5.8	− 7.6	− 9.4	−11.2	−13.0	−14.8	−16.6	−18.4	−20.2
−30	−22.0	−23.8	−25.6	−27.4	−29.2	−31.0	−32.8	−34.6	−36.4	−38.2
−40	−40.0	−41.8	−43.6	−45.4	−47.2	−49.0	−50.8	−52.6	−54.4	−56.2

Note: The vertical axis lists degree Centigrade in ten-degree increments; the horizontal axis lists degrees Centigrade in one-degree increments. For example, to convert −23 degrees Centigrade to Fahrenheit, read down the vertical axis to −20 and across the horizontal axis to 3. The answer in this case is −9.4 degrees Fahrenheit.

Table A–8. Temperature Differences: Degrees Centigrade to Fahrenheit.

Degrees Centigrade	Degrees Fahrenheit	Degrees Centigrade	Degrees Fahrenheit
1	1.8	6	10.8
2	3.6	7	12.6
3	5.4	8	14.4
4	7.2	9	16.2
5	9.0	10	18.0

Note: This table indicates the number of degrees Fahrenheit the temperature will rise for each increase in degrees Centigrade. For instance, if the temperature increases 7 degrees Centigrade, the comparable temperature in degrees Fahrenheit will increase 12.6 degrees. (A temperature increase from 10°C to 17°C equals an increase from 50° F to 62.6°F.)

Table A–9(a). Inches of Mercury (in round numbers) in Millibars.

Inches of Mercury	Millibars	Inches of Mercury	Millibars	Inches of Mercury	Millibars
1	33.86	11	372.50	21	711.14
2	67.73	12	406.37	22	745.01
3	101.59	13	440.23	23	778.87
4	135.46	14	474.09	24	812.73
5	169.32	15	507.96	25	846.60
6	203.18	16	541.82	26	880.46
7	237.05	17	575.69	27	914.33
8	270.91	18	609.55	28	948.19
9	304.78	19	643.41	29	982.05
10	338.64	20	677.28	30	1015.92
				31	1049.78
				32	1083.65

Table A–9(b). Inches of Mercury (in tenths and hundedths) to Millibars.

Inches of Mercury	0.00	.01	.02	.03	.04	.05	.06	.07	.08	.09
0.0	0.00	0.34	0.68	1.02	1.35.	1.69	2.03	2.37	2.71	3.05
0.1	3.39	3.73	4.06	4.40	4.74	5.08	5.42	5.76	6.10	6.43
0.2	6.77	7.11	7.45	7.79	8.13	8.47	8.80	9.14	9.48	9.82
0.3	10.16	10.50	10.84	11.18	11.51	11.85	12.19	12.53	12.87	13.21
0.4	13.55	13.88	14.22	14.56	14.90	15.24	15.58	15.92	16.25	16.59
0.5	16.93	17.27	17.61	17.95	18.29	18.63	18.96	19.30	19.64	19.98
0.6	20.32	20.66	21.00	21.33	21.67	22.01	22.35	22.69	23.03	23.37
0.7	23.70	24.04	24.38	24.72	25.06	25.40	25.74	26.08	26.41	26.75
0.8	27.09	27.43	27.77	28.11	28.45	28.78	29.12	29.46	29.80	30.14
0.9	30.48	30.82	31.15	31.49	31.83	32.17	32.51	32.85	33.19	33.53

Table A–9(a) converts inches of mercury in round numbers to millibars. Table A–9(b) converts inches of mercury in tenths and hundredths to millibars. Both tables can be used together to solve a conversion problem.

Example: Convert 29.73 inches of mercury to millibars.

Solution: First determine from table A–9(a) the number of millibars in 29 inches of mercury.

Then determine from table A–9(b) the number of millibars in .73 inches of mercury. In table A–9(b), the vertical axis lists inches of mercury in tenths of an inch. The horizontal axis lists inches of mercury in hundredths of an inch. To convert .73 inches, read down the vertical axis to .7 and across the horizontal axis to .03.

Now add the data from both tables for the answer.

From table A–9(a)   982.05 mbs
From table A–9(b)   24.72 mbs

   Answer: 1006.77 mbs

Table A–10. Begin to think "Metric." The International Metric System is founded on six base units of measurement and all units are based on multiples of 10. Standard prefixes (such as milli, centi, deci, deca, hecto, and kilo) are added to the base units to give names for quantities of a particular unit that differ by multiples of 10. For example, a millimeter is one thousandth of a meter (0.001 m). A kilometer is one thousand meters (1,000 m).

Metric Conversion Factors

1 centimeter = 10 millimeters
1 decimeter = 10 centimeters
1 meter = 10 decimeters
1 meter = 100 centimeters
1 decameter = 10 meters
1 hectometer = 10 decameters
1 kilometer = 10 hectometers
1 kilometer = 1,000 meters

Metric vs American

Length

1 inch = 2.54 centimeters
1 foot = 30.48 centimeters
1 yard = 91.44 centimeters
1 meter = 100.00 centimeters
1 meter = 39.37 inches
1 meter = 3.28 feet
1 meter = 1.09 yards
1 kilometer = 0.621 miles
1 mile = 1.609 kilometers

Weight

1 pound = 0.454 kilograms
1 kilogram = 2.2 pounds
1 kilogram = 1,000.0 grams

Volume

1 cubic yard = 0.765 cubic meters
1 cubic meter = 35.31 cubic feet
1 cubic meter = 1.308 cubic yards

Liquid

1 gallon = 3.785 liters
1 liter = 0.264 gallons

Power

1 kilowatt = 1,000.0 watts
1 horsepower = 0.746 kilowatts
1 kilowatt = 1.34 horsepower

APPENDIX B. RECOMMENDED BOOKS AND PERIODICALS

Books

Conant, James B. *On Understanding Science.* New York: The New American Library, 1951.

Douglas, Marjory S. *Hurricane.* New York: Rinehart and Co., 1958.

Dunn, G. E. and Miller, B. I. *Atlantic Hurricanes.* Baton Rouge: Louisiana State University Press, 1960.

Harding, E. T. and Kotsch, W. J. *Heavy Weather Guide.* Annapolis: U. S. Naval Institute, 1965.

Hess, S. L. *Introduction to Theoretical Meteorology.* New York: Holt, Rinehart and Winston, Inc., 1959.

Huschke, R. E., ed. *Glossary of Meteorology.* Boston: American Meteorological Society, 1959.

Kendrew, W. G. *The Climates of the Continents,* 5th ed. Fair Lawn, N.J.: Oxford University Press, 1961.

Mariner's Worldwide Climate Guide to Tropical Storms at Sea. Washington: U. S. Government Printing Office, 1974.

Meteorology for Mariners, Met. O. 593, 2nd ed. London: Her Majesty's Stationery Office, 1967.

Middleton, W.E.K. and Spilhaus, A.F. *Meteorological Instruments,* 3rd rev. ed. Toronto: University of Toronto Press, 1953.

Neuberger, H. *Introduction to Physical Meteorology.* University Park: The Pennsylvania State University Press, 1951.

Neuberger, H. and Stevens, F. B. *Weather and Man.* Englewood Cliffs: Prentice-Hall, Inc., 1948.

Noel, Capt. J. V., Jr., ed. *Knight's Modern Seamanship.* New York: D. Van Nostrand, 1966.

Panofsky, H. *Introduction to Dynamic Meteorology.* University Park: The Pennsylvania State University Press, 1956.

Reiter, E. R. *Jet Streams.* Garden City: Doubleday and Co., Inc., 1967.

Riehl, H. *Introduction to the Atmosphere.* New York: McGraw-Hill Book Co., Inc., 1965.

Smithsonian Meteorological Tables. Washington: The Smithsonian Institution, 1951.

Stewart, G. R. *Storm.* New York: Modern Library, Inc., 1947.

Sverdrup, H. U. *Oceanography for Meteorologists.* Englewood Cliffs: Prentice-Hall, Inc., 1942.

Tobin, Wallace E. III. *The Mariner's Pocket Companion.* Annapolis: U. S. Naval Institute, 1975.

Trewartha, G. T. *An Introduction to Climate,* 3rd ed. New York: McGraw-Hill Book Co., Inc., 1954.

U. S. Navy. *Marine Climatic Atlas of the World,* volumes on the Atlantic, Pacific, and Indian Oceans. Washington: Government Printing Office, published beginning 1955.

Willett, H. C. and Sanders, F. *Descriptive Meteorology,* 2nd ed. New York: Academic Press, Inc., 1959.

Periodicals

Average Monthly Weather Resume and Outlook. Semi-monthly. Washington: NOAA National Weather Service.

Bulletin of the American Meteorological Society. Monthly. Boston: American Meteorological Society.

Bulletin of the World Meteorological Organization. Quarterly. Geneva, Switzerland: World Meteorological Society.

Journal of Applied Meteorology. Monthly. Boston: American Meteorological Society.

Mariner's Weather Log. Monthly. Washington: NOAA National Weather Service.

Monthly Weather Review. Monthly. Washington: Government Printing Office.

Weatherwise. Bimonthly. Boston: American Meterological Society.

APPENDIX C. HURRICANE AND TYPHOON NAMES

Hurricanes–Eastern North Pacific (East of 140° west longitude)

1974	1975	1976	1977
Aletta	Agatha	Annette	Ava
Blanca	Bridget	Bonny	Bernice
Connie	Carlotta	Celeste	Claudia
Dolores	Denise	Diana	Doreen
Eileen	Eleanor	Estelle	Emily
Francesca	Francene	Fernanda	Florence
Gretchen	Georgette	Gwen	Glenda
Helga	Hilary	Hyacinth	Heather
Ione	Ilsa	Iva	Irah
Joyce	Jewel	Joanne	Jennifer
Kristen	Katrina	Kathleen	Katherine
Lorraine	Lily	Liza	Lillian
Maggie	Monica	Madeline	Mona
Norma	Nanette	Naomi	Natalie
Orlene	Olivia	Orla	Odessa
Patricia	Priscilla	Pauline	Prudence
Rosalie	Ramona	Rebecca	Roslyn
Selma	Sharon	Simone	Sylvia
Toni	Terry	Tara	Tillie
Vivian	Veronica	Valerie	Victoria
Winona	Winifred	Willa	Wallie

Hurricanes–Central North Pacific (from 140° west longitude westward to the 180th meridian).

1974	1975	1976	1977
Alice	Anita	Amy	Agnes
Betty	Billie	Babe	Bess
Cora	Clara	Carla	Carmen
Doris	Dot	Dinah	Della
Elsie	Ellen	Emma	Elaine
Flossie	Fran	Freda	Faye
Grace	Georgia	Gilda	Gloria
Helen	Hope	Harriet	Hester
Ida	Iris	Ivy	Irma
June	Joan	Jean	Judy
Kathy	Kate	Kim	Kit
Lorna	Louise	Lucy	Lola
Marie	Marge	Mary	Mamie
Nancy	Nora	Nadine	Nina
Olga	Opal	Olive	Ora
Pamela	Patsy	Polly	Phyllis
Ruby	Ruth	Rose	Rita
Sally	Sarah	Shirley	Susan
Therese	Thelma	Trix	Tess
Violet	Vera	Virginia	Viola
Wilda	Wanda	Wendy	Winnie

Hurricanes—Atlantic, Caribbean and Gulf of Mexico

1974	1975	1976	1977	1978	1979	1980	1981	1982	1983
Alma	Amy	Anna	Anita	Amelia	Angie	Abby	Arlene	Agnes	Alice
Becky	Blanche	Belle	Babe	Bess	Barbara	Bertha	Beth	Betty	Brenda
Carmen	Caroline	Candice	Clara	Cora	Cindy	Candy	Chloe	Carrie	Christine
Dolly	Doris	Dottie	Dorothy	Debra	Dot	Dinah	Doria	Dawn	Delia
Elaine	Eloise	Emmy	Evelyn	Ella	Eve	Elsie	Edith	Edna	Ellen
Fifi	Faye	Frances	Frieda	Flossie	Franny	Felicia	Fern	Felice	Fran
Gertrude	Gladys	Gloria	Grace	Greta	Gwyn	Georgia	Ginger	Gerda	Gilda
Hester	Hallie	Holly	Hannah	Hope	Hedda	Hedy	Heidi	Harriet	Helen
Ivy	Ingrid	Inga	Ida	Irma	Iris	Isabel	Irene	Ilene	Imogene
Justine	Julia	Jill	Jodie	Juliet	Judy	June	Janice	Jane	Joy
Kathy	Kitty	Kay	Kristina	Kendra	Karen	Kim	Kristy	Kara	Kate
Linda	Lilly	Lilias	Lois	Louise	Lana	Lucy	Laura	Lucille	Loretta
Marsha	Mabel	Maria	Mary	Martha	Molly	Millie	Margo	Mae	Madge
Nelly	Niki	Nola	Nora	Noreen	Nita	Nina	Nona	Nadine	Nancy
Olga	Opal	Orpha	Odel	Ora	Ophelia	Olive	Orchid	Odette	Ona
Pearl	Peggy	Pamela	Penny	Paula	Patty	Phyllis	Portia	Polly	Patsy
Roxanne	Ruby	Ruth	Raquel	Rosalie	Roberta	Rosie	Rachel	Rita	Rose
Sabrina	Sheila	Shirley	Sophia	Susan	Sherry	Suzy	Sandra	Sarah	Sally
Thelma	Tilda	Trixie	Trudy	Tanya	Tess	Theda	Terese	Tina	Tam
Viola	Vicky	Vilda	Virginia	Vanessa	Vesta	Violet	Verna	Velma	Vera
Wilma	Winnie	Wynne	Willene	Wanda	Wenda	Willette	Wallis	Wendy	Wilda

Typhoons—Western North Pacific (west of the 180th meridian).

1974	1975	1976	1977
Alice	Anita	Amy	Agnes
Betty	Billie	Babe	Bess
Cora	Clara	Carla	Carmen
Doris	Dot	Dinah	Della
Elsie	Ellen	Emma	Elaine
Flossie	Fran	Freda	Faye
Grace	Georgia	Gilda	Gloria
Helen	Hope	Harriet	Hester
Ida	Iris	Ivy	Irma
June	Joan	Jean	Judy
Kathy	Kate	Kim	Kit
Lorna	Louise	Lucy	Lola
Marie	Marge	Mary	Mamie
Nancy	Nora	Nadine	Nina
Olga	Opal	Olive	Ora
Pamela	Patsy	Polly	Phyllis
Ruby	Ruth	Rose	Rita
Sally	Sarah	Shirley	Susan
Therese	Thelma	Trix	Tess
Violet	Vera	Virginia	Viola
Wilda	Wanda	Wendy	Winnie

index

The text of this book is set in ten-point Caledonia with two points of leading. The chapter titles are thirty-six-point Friz Quadrata.

The book is printed offset on Finch Opaque Smooth paper. The cover is Joanna Devon 40800, linen finish.

Edited by Carol Swartz.

Designed by Beverly Baum.

The book was composed by Modern Typographers, Inc., Clearwater, Florida.

The book was printed by The John D. Lucas Printing Company, Baltimore, Maryland, and bound by The Delmar Company, Charlotte, North Carolina.

PHOTOGRAPH CREDITS

R.K. Pilsbury, ii, 48, 180; Jan Hahn, 2; NASA, 26, 80; U.S. Coast Guard, 66, 204; U.S. Navy, 98, 194, 254; Toby Marquez, 118; NOAA National Environmental Satellite Service, 136,203.